"AI is often discussed in terms of technology, but this book focuses on what matters most, opportunity. Through insights from 33 experts across multiple industries, The AI Opportunity *demonstrates how artificial intelligence is already transforming work, business, healthcare, education, and daily life. An inspiring and practical guide for anyone who wants to thrive in the age of AI."*

—MICHAEL RICHARDS, PhD, AI Researcher, Keynote Speaker

"Rather than viewing artificial intelligence as something to fear, The AI Opportunity *invites readers to see it as one of the greatest opportunities of our generation. With diverse perspectives from global experts, this book provides a compelling roadmap for understanding AI's impact and positioning yourself for success in a rapidly changing world."*

—DAVID BROWNLEE, Founder of The Brownlee Group, LinkedIn Learning Instructor

"In a world filled with headlines about AI, few books successfully connect technological innovation with real-world human potential. The AI Opportunity *does exactly that. By showcasing how AI is being applied across industries and professions, this book empowers readers to move from curiosity to confidence and from uncertainty to action."*

—SHAWN FECHTER, Sales & Marketing Executive

"The AI Opportunity *is a timely reminder that every major technological shift creates winners and losers, not because of luck, but because of awareness and action. This book provides the knowledge and perspective needed to recognize the opportunities AI presents and use them to build a better future."*

—SARAH LIDEL, CEO of AIworks

"One of the greatest strengths of this book is its diversity of voices. Bringing together 33 experts from around the world, The AI Opportunity *offers a balanced, accessible, and highly practical exploration of how artificial intelligence is reshaping civilization. This is essential reading for professionals, entrepreneurs, educators, and lifelong learners alike."*

—DONALD CHEN, Lecturer at Innovation Nexus

"Artificial intelligence is rapidly becoming woven into the fabric of modern life, yet many people still struggle to understand where they fit into this transformation. The AI Opportunity *bridges that gap beautifully. Insightful and encouraging, this book helps readers see not only where AI is headed, but how they can benefit from it today."*

—REBECCA HARLEY, Author, Futurist, AI Advocate

THE AI OPPORTUNITY

THE AI OPPORTUNITY

Unlocking Potential Within Civilization's Next Big Disruption

Authored by:
Erik Seversen, Gabe Arrington, Saptarshi Basu, Nati Berkover, Phani Burra, Chris Clarkson, Alex Dean, Eric Dressler, Barry Flanagan , Stephen Gatchell, T. Alexander Gorrie, Christopher Hartline, Stephan Heilmann, Anant Kalyan, Gilad Langer, PhD, Christopher Lomas, Sriya Burra, Maribeth Martorana, Leonardo Veri, Rafa Maximo, Vinodhini Ravikumar, Harvey Schinkal, Jasmin Jijina Sinkular, PhD, Sylvia Stocker, Sakina Syed, Vishruth Burra, Daniel Jonathan Valik, Hardik Modi, Bobbie Wasserman, Cindy Weber, Christi Wharton, Eddie James Williams III, Pang Sern Yong, Svetlana Zavelskaya, and Steven Yiliang Zhou

THIN LEAF PRESS | LOS ANGELES

Library of Congress Cataloging-in-Publication Data
Names: Seversen, Erik, Author, et al.
Title: *The AI Opportunity: Unlocking Potential Within Civilization's Next Big Disruption*
LCCN: 2026914434

ISBN 978-1-968318-66-6 (hardcover) | 978-1-968318-65-9 (paperback)
ISBN 978-1-968318-64-2 (eBook) | 978-1-968318-67-3 (audiobook)

Artificial Intelligence, Science & Technology, Business, Professional Development
Cover Design: 100 Covers
Interior Design: Dindo Sanguenza
Editor: Nancy Pile
Thin Leaf Press
Los Angeles

THIN
LEAF

Thank you for reading this book. There is information found within the following pages that can greatly benefit your life, but don't stop there. Make sure you get the most you can from this book and reach out directly to the expert-authors who want to help you navigate the AI opportunity, so you can thrive within civilization's next big disruption, and manifest success in your life. Contact information for each author is found at the end of their respective chapter.

To those at the frontlines of technology seeking ways to make the world a better place for all.

CONTENTS

INTRODUCTION xix
By Erik Seversen
Author of *Ordinary to Extraordinary*
Los Angeles, California

CHAPTER 1
AI, NATIONAL SECURITY, AND THE
ARCHITECTURE OF GLOBAL ORDER..1
By Gabe Arrington
Colonel, United States Air Force
Austin, Texas

CHAPTER 2
FROM MACHINES TO LIVING SYSTEMS: LEADING
THROUGH THE AUTOMOTIVE SHIFT 13
By Saptarshi Basu
Techpreneur championing the Graypreneur economy
Singapore

CHAPTER 3
EVERYONE BECOMES A BUILDER 25
By Nati Berkover
AI Builder, Entrepreneur, and Product Leader
Tel Aviv, Israel

CHAPTER 4
THE AI INDUSTRIAL RESET: BUILDING FACTORIES OF INTELLIGENCE 35
By Phani Burra
VP, AI & Digital; Architecting AI-First Enterprises
Dublin, California

CHAPTER 5
ARCHITECTING TRUST AND ENGINEERING TRUTH.......... 43
By Chris Clarkson
AI Technologist and Consultant
Adelaide, Australia

CHAPTER 6
HUMAN JUDGEMENT, AI POWER, AND THE RACE FOR OPERATIONAL EXCELLENCE 51
By Alex Dean
AI Advisor and CIO Whisperer
Zurich, Switzerland

CHAPTER 7
THE STICKY NOTE THAT CHANGED EVERYTHING............ 61
By Eric Dressler
AI Leader and Founder, Agentsium
Edmonton, Canada

CHAPTER 8
THE EVOLUTION OF B2B SELLERS IN THE AGE OF AI......... 71
By Barry Flanagan
Founder, AI-IQ Systems
Granger, Indiana

CHAPTER 9
FROM DATA TO DECISIONS: MAKING AI WORK FOR REAL BUSINESS VALUE.. 81
By Stephen Gatchell
Partner and Head of AI Strategy
Boston, Massachusetts

CHAPTER 10
EVERYONE HAD A PLAN 95
By T. Alexander Gorrie
Founder and CEO
London, England, United Kingdom

CHAPTER 11
THE AMPLIFIED HUMAN 101
By Christopher Hartline
AI & Cloud Architect
Lenexa, Kansas

CHAPTER 12
AI WON'T REPLACE YOU, BUT PEOPLE WHO USE
IT WILL 111
By Stephan Heilmann
AI Leader and Advocate
Zug, Switzerland

CHAPTER 13
GONE IN 60 SECONDS: THE ENTERPRISE THAT
LEARNED TO HEAL ITSELF 123
By Anant Kalyan
Founder & CEO, Alloi.ai
Atlanta, Georgia

CHAPTER 14
DESIGNING AI-ENABLED OPERATIONS THAT
HUMANS CAN TRUST 137
By Gilad Langer, PhD
Composable Operations Strategist
Castro Valley, California

CHAPTER 15
THE RISE OF ENTERPRISE (ARTIFICIAL) INTELLIGENCE ..153
By Christopher Lomas
SVP, UK Digital & Product Management Leader
London, England, United Kingdom

CHAPTER 16
FROM SCREENS TO REALITY: AI, AR/VR, AND THE RISE OF EMBODIED INTELLIGENCE 167
By Sriya Burra
AI Researcher and Emerging Technology Innovator
Dublin, California

CHAPTER 17
HOW C-SUITE LEADERS CAN BUILD AI THAT ACTUALLY DELIVERS ROI 175
By Maribeth Martorana, CFA
Founder & CEO, Sapient Advisors: AI, Innovation
Miami, Florida

CHAPTER 18
CLARITY IS THE REAL AI OPPORTUNITY 185
By Leonardo Veri
Founder, CTO, and AI Advisor
São Paulo, Brazil

CHAPTER 19
THE HUMAN LINCHPIN: THE BETTER AI GETS, THE MORE VALUABLE YOU BECOME 195
By Rafa Maximo
AI Marketing; Creator of The Human Linchpin System
Miami, Florida

CHAPTER 20
EMBODIED INTELLIGENCE: THE RISE OF PHYSICAL AI 205
By Vinodhini Ravikumar
AI Leader and Co-Founder at Vault Voyage Ltd
Seattle, Washington

CHAPTER 21
FROM CAPACITY TO CONSCIENCE: THE REAL WORK OF ORGANIZATIONAL MATURITY 217
By Harvey Schinkal
Analytics Leader, Doctoral Candidate, Purdue University
Houston, Texas

CHAPTER 22
By Jasmin Jijina Sinkular, PhD
Technology & Sustainability Innovator & Consultant
Detroit, Michigan

CHAPTER 23
HUMANS AND ROBOTS: AMPLIFYING LEADERSHIP IN THE AGE OF AI 243
By Sylvia Stocker
Global Robotics & AI Thought Leader
Zurich, Switzerland

CHAPTER 24
SCALING LAWS: THE STRANGE PHYSICS OF INTELLIGENCE 255
By Sakina Syed
Microsoft IT Professional, AI Engineer, and Consultant
Toronto, Canada

CHAPTER 25
QUANTUM COMPUTING AND THE FUTURE OF ECONOMIC DECISION-MAKING 271
By Vishruth Burra
DECA Award Winner; AI & Quantum Finance Researcher
Dublin, California

CHAPTER 26
FROM CONNECTIVITY TO AUTONOMY: THE FUTURE OF HUMAN COLLABORATION AND AGENTIC AI 277
By Daniel Jonathan Valik
Technology & Product Executive; AI, Data Platforms, CX
Seattle, Washington
By Hardik Modi
Product Leader, AI, Data, & Marketing Tech Platforms
San Francisco, California

CHAPTER 27
THE MISSING INTELLIGENCE LAYER: 293
HOW AI WILL HELP HUMANS NAVIGATE
COMPLEX SYSTEMS....................................293
By Bobbie Wasserman, MBA
Founder, The Homeownership Collective; Civic Systems
Los Angeles, California

CHAPTER 28
AI IN LIFE SCIENCES: A JOURNEY OF
TRANSFORMATION....................................303
By Cindy Weber
AI Solutions Leader
Medford, Massachusetts

CHAPTER 29
THE WISDOM GAP: WHY DISCERNMENT IS YOUR
AI COMPETITIVE ADVANTAGE 315
By Christi Wharton
Mental Health and Professional Life Coach
Baton Rouge, Louisiana

CHAPTER 30
GUARDRAILS AND GAS PEDALS: GOVERNING AI
AS A BUSINESS DRIVER IN AN AGE OF RISK....................................325
By Eddie James Williams III
Head of Risk & Compliance (Deputy CISO)
Kansas City, Missouri

CHAPTER 31
THE NEW SCARCITY IS TASTE 339
By Pang Sern Yong
AI Strategist & Coach
Kuala Lumpur, Malaysia | Singapore

CHAPTER 32
BUILDING AI YOU CAN TRUST: LESSONS FROM A RISK-AVERSE INDUSTRIES...349
By Svetlana Zavelskaya, MS, MBA
Senior Technology Leader
New York, New York

CHAPTER 33
THE PERFECT EXCUSE: WHO IS ACCOUNTABLE FOR AI?.361
By Steven Yiliang Zhou
AI Transformation & Cybersecurity Expert
San Francisco, California

DID YOU ENJOY THIS BOOK?...371

INTRODUCTION

By Erik Seversen
Author of *Ordinary to Extraordinary*
Los Angeles, California

Artificial Intelligence is the biggest commercial opportunity for companies, industries, and nations over the next few decades.
—PwC report 2017

At any given time there exist opportunities in front of us. In my life, one of the first opportunities (that I'm aware of) occurred when I was five years old and my elderly neighbor, Mr. Wesley, plowed about a 10x10 ft square in our field and gave me some carrot and onion seeds. I think he mainly did this to get me to quit following him around his massive vegetable garden all day, so I'd have something to occupy my time.

As soon as I saw the first sprouts of onions coming up, I knew there would be way more than enough for my family to eat, so I decided to sell them on the side of the road. This was both my first entrepreneurial adventure, and it was also the first time I saw an opportunity and took advantage of it. I tended the garden and, yes, I sold all the onions for two dollars, which was a massive amount of money for a five-year-old in 1974. The point of the story is that Mr. Wesley gave me some seeds to keep me busy, but I recognized the opportunity that was there and capitalized on it.

As I got a bit older, at about 12 years old, I was the primary lawn mower at my house since my dad worked most Saturdays and my teenage brother and sister were occupied with team sports and activities. When I decided I wanted a bit more spending money, I knocked on every door on the block and asked if I could mow their lawn. This turned into a weekly moving routine of three different houses in my neighborhood at between $10 and $20 each. Again, I was doing pretty well for young kid. And again, this was only due to me seeing an opportunity. No one told me that moving other people's yards was a thing, but I saw something when there were no instructions and no signs, and again I capitalized on it.

The last thing I'll mention was when I was just out of graduate school. I was an avid motorcycle rider, and I had also done a ton of traveling by that time to see family internationally, for my anthropology studies, and for different jobs I'd had abroad. I knew two things: I loved riding motorcycle and I loved meeting people from other countries, So I created a company called RideFree Motorcycle Tours. With this company, I brought people from around the world to ride motorcycles with me. I got to ride through some of the most beautiful landscapes in the USA, I met a ton of wonderful international people, and I made money while doing it. Yep, I saw an opportunity and used it to make my life better.

You see, opportunities are around us all of the time, but it is up to us to capitalize on them. While selling onions, mowing lawns, and creating a travel company were opportunities for me based on my skills and passions, there is an opportunity in front of all of us right now. Yes, you guessed right. That opportunity is AI.

I've been a fan of AI for a while now, but I wasn't an early adopter. I had been hearing about bits and pieces of AI, but it wasn't until ChatGPT put the spotlight on GenAI and Large Language Models that I began to get excited about what was happening with this new technology. For quite a while, it seemed many people considered ChatGPT to be the sum of AI. Now, with the advent of multiple GenAI platforms including Gemini, Claude, Copilot, Perplexity, Grok, and more, we have numerous choices. We're also now seeing AI in predictive analytics (finance, healthcare), computer vision (healthcare, autonomous driving), natural language models (customer service chatbots, translation), robotics (manufacturing, distribution), fraud detection (cybersecurity, e-commerce), optimization (scheduling, logistics), and much more. AI is essentially becoming part of the fabric in which we live. While no one can really escape AI, quite a few people are working and living as they always have. This is fine for now, but it won't be long before people begin to fall into two camps: Those who are using AI intelligently and those who aren't.

The goal of this book is to help people see the AI opportunity that is staring all of us in the face and to help people who are smart and educated but who don't know a lot about AI to see that using AI intelligently will create an advantage for those who use it while those who avoid AI will, at some point, fall behind.

As a writer and educator, it is my goal to help as many people as possible become familiar enough with AI that they will be among those who thrive. Since I'm not an AI expert, I didn't try to write this book about AI by myself. Rather I solicited the help of 33 AI experts from various backgrounds and locations. The co-authors of this book come from all over the USA, Canada, The United Kingdom, Germany, Switzerland, Israel, Malasia, Singapore, and Australia.

These authors have been featured on NBC, FOX, ABC, CBS, BBC, CSUN, Sky News, and more. They are professionals who are university professors, business owners, consultants, product managers, engineers, software developers, cybersecurity leaders, healthcare experts, advisors, TEDx and keynote speakers, researchers, platform architects, senior data analysts, IT directors, CTOs, and more. The one thing these individuals have in common is that they all have something to share about artificial intelligence, and these ideas are available to you now.

Although this book is organized around the united theme of using AI as an opportunity within civilization's next big disruption, each of the chapters is totally stand-alone. The chapters in the book can be read in any order. I encourage you to look through the table of contents and begin wherever you want. However, I urge you to read all the chapters because, as a whole, they provide a great array of perspectives. Each is valuable in helping you understand how AI is being used in many areas of human industry and human life.

It is my hope that you discover something in this book that helps you navigate civilization's next big disruption as humans and machines become more interdependent and AI technology continues to insert itself into many aspects of our lives. It is my hope that you both embrace AI as you take advantage of the AI opportunity as humans and machines work together to create a positive future for all.

About the Author

Erik Seversen is on a mission to inspire people. He holds a master's degree in anthropology and is a certified practitioner of neuro-linguistic programming. Erik draws from his years of teaching at the university level and years of real-life experience in business to motivate people to take action, creating extreme success in business and in life.

Erik is a TEDx and keynote speaker who has reached over one million people through his public speaking and live courses. He has visited over 100 countries and all 50 states in the USA and has climbed the highest mountains on four continents, 15 countries, and 18 states. Erik has published 22 bestselling books on the topics of mindset, success, and peak performance, and he has helped over 500 people become authors. He is a full-time writer, book consultant, and speaker, and he lives by the idea that success is available to everyone—that living an extraordinary life is a choice.

Erik lives in Los Angeles with his wife and has two boys currently studying at university.

Contact Erik for interviews, speaking, or book publishing consultation.

Email: Erik@ErikSeversen.com

Website: www.ErikSeversen.com

LinkedIn: https://www.linkedin.com/in/erikseversen/

CHAPTER 1

AI, NATIONAL SECURITY, AND THE ARCHITECTURE OF GLOBAL ORDER

By Gabe Arrington
Colonel, United States Air Force
Austin, Texas

Whoever leads in artificial intelligence will rule the world.
—Vladimir Putin, September 1, 2017

Vladimir Putin did not say those words to comfort the West or Westphalian order. He said them as a declaration of intent for the future of the international world order. In the years since, that declaration has matured into doctrine and policy, not just in Moscow, but in Beijing, Brussels, and Washington. Artificial intelligence is no longer a technological story. It is a power story. As with every contest for power in modern history, it is playing out diplomatically, informationally, militarily, and economically simultaneously.

This chapter makes a straightforward proposal: AI is the connective tissue of modern national power. It is not a single weapon, a single tool, or a single policy challenge. It is the path through which every domain of national power—military readiness, industrial capacity, intelligence collection, diplomatic leverage, and statecraft itself—will increasingly flow. Nations that understand this early will shape the architecture of the global order. Nations that fail to act will find themselves operating inside an international order designed by state competitors.

The United States sits at the center of this contest, but sitting at the center is not the same as winning it. Advantage in AI is not guaranteed by history, by democracy, or by market capitalism alone. It must be built, sustained, and governed, deliberately and urgently.

The Strategic Landscape: AI Across Four Instruments of National Power

To understand AI's national security implications, it helps to think in four pillars of national power: diplomatic, informational, military, and economic. Each represents a distinct domain of competition. Together, they form the complete picture of why AI has elevated to the top of every serious security agenda worldwide.

In the diplomatic pillar AI is quietly restructuring the levers of statecraft itself. Foreign policy has always depended on information advantage: knowing more than your counterpart, understanding the second- and third-order consequences of an action before committing to it, reading the political dynamics of a partner nation with accuracy and speed. We see this play out on the world stage daily, documented by the 24-hour news cycle. AI amplifies each of these capabilities, and it does so asymmetrically. Nations with mature AI-enabled diplomatic infrastructure can process treaty compliance signals, model economic coercion scenarios, interpret nuclear ambitions, and anticipate alliance fractures in near real-time. Nations without it are negotiating with one hand behind their back.

AI is also reshaping the tools of soft power. The ability to target foreign populations with precisely calibrated messaging—in their language, tuned to their cultural context, timed to moments of political vulnerability—represents a qualitative leap in global influence operations. China's use of AI-enabled content generation across Southeast Asian and African digital media markets is not speculative but documented.[1] Russia's interference operations have similarly evolved from blunt-instrument trolling to AI-assisted micro-targeting.[2] The diplomatic implications are significant: When the information environment in a partner nation is being actively shaped by an adversary's AI systems, traditional diplomacy—built on shared facts and trusted channels—operates in a compromised setting.

Perhaps most consequentially, AI is beginning to affect the internal deliberations of governments themselves. Machine-assisted policy analysis, automated red-teaming of diplomatic options, and AI-generated scenario modeling are entering the decision support systems of major powers. This is not inherently destabilizing; better analysis can produce better decisions.[3] It does, however, introduce new risks around over-reliance, algorithmic bias embedded in geopolitical recommendations, and the speed compression of diplomatic timelines. Crises that once allowed hours or days for careful deliberation, such as the Cuban Missile Crisis, may compress to minutes when AI-enabled decision support is advising policymakers on both sides simultaneously. The diplomatic community has not yet fully reckoned with this reality.

In the informational pillar, AI has fundamentally altered the intelligence environment, sometimes leading to a question of what truth is. Large-scale open-source intelligence collection, natural language processing across foreign-language media, pattern recognition in satellite imagery—all of these have been supercharged by AI.[4] Simultaneously, AI has lowered the cost and raised the sophistication of influence operations, disinformation campaigns, and cyber intrusions. The information domain, already contested, is now dramatically more dangerous. The use of widespread data through AI is immediate, often without fail-safe options to ensure data integrity.

In the military pillar, AI is reshaping the character of warfare. Autonomous systems—aerial, maritime, and ground-based—are already deployed in operational environments. Ukraine's use of AI-enabled drone swarms, real-time battlefield data fusion, and machine-assisted targeting represents not a glimpse of future war, but a live demonstration of its present form.[5] China's use of satellite imagery and AI-enabled targeting allowed Iran to perform kinetic strikes across the Middle East.[6] AI enables compression of the kill chain: the sequence from sensing a threat to responding to it. What once required hours of analyst review can now occur in seconds. Speed asymmetry in warfare is not marginal. It is decisive.

The People's Liberation Army has made AI-enabled "intelligentized warfare" an explicit strategic priority.[7] China's Military-Civil Fusion doctrine deliberately eliminates the firewall between commercial AI development and defense application. A Chinese company building logistics optimization software is, by policy design, also building capability that can be redirected toward military supply chain management. This is not conspiracy theory. It is written into Chinese law.

In the economic pillar, AI is restructuring industrial power in ways that compound over time. Manufacturing, aerospace, and logistics—the physical strength of military capability—are all undergoing AI-driven transformation. Predictive maintenance extends the operational life of aircraft and vehicles. AI-enabled supply chain management reduces vulnerability to disruption. Autonomous manufacturing compresses production timelines. Nations that lead in AI-driven industrial productivity will have deeper and faster arsenals. This is not an abstraction; it is a direct line from algorithm to airframe.

The American Position: Strengths, Vulnerabilities, and the Risk of Complacency

The United States retains meaningful advantages in AI. American universities produce a disproportionate share of the world's top AI researchers.[8] The US venture capital ecosystem funds AI development at a scale no other nation matches.[9] American technology companies—developing the foundational models, the chips, and the cloud infrastructure that underpin modern AI—represent a genuine structural advantage.

But advantages are not permanent, and several trends warrant serious concern.

First, the integration gap. The US defense enterprise has historically been slow to absorb commercial technology. The acquisition system, designed for the deliberate procurement of exquisite hardware, is poorly calibrated to the rapid iteration cycles of software and AI. Programs that take seven years to field are irrelevant in a technology domain where capability doubles in 18 months. The Department of War has taken meaningful steps—the creation of the Chief Digital and AI Office (CDAO), the Replicator Initiative, the proliferation of software factories—but the pace of institutional adaptation still lags the pace of technological change. In a system designed to spend taxpayer money slowly, AI incorporation demands systemic change for rapid adoption across the defense industrial base.

Second, the talent pipeline. AI expertise is globally mobile and privately compensated. Competing for AI talent against technology companies that offer stock options, flexible culture, and creative latitude is a structural challenge for the federal government. Classification requirements, security clearance timelines, and a bloated bureaucratic culture create friction where private industry does not. The US government must find creative mechanisms—reserve programs, rotational assignments, public-private partnerships—to retain access to the talent it needs.

Third, the dependency problem. The semiconductor supply chain, upon which all AI capability ultimately rests, remains concentrated in ways that create strategic risk. Taiwan Semiconductor Manufacturing Company (TSMC) produces most the world's most advanced chips.[10] The CHIPS and Science Act represent a serious attempt to reshore advanced semiconductor production, but fabs take years to build and decades to mature. The US must accelerate organic chip production as a matter of strategic importance and vital to its national security. The vulnerability exists today. The mitigation is years away.

The Governance Imperative: Norms, Treaties, and the Problem of Lethal Autonomy

Technology without governance is simply acceleration toward unknown outcomes. The history of weapons development offers instructive parallels. Biological weapons, chemical weapons, and nuclear arms all required international frameworks—imperfect, sometimes violated, but consequential nonetheless—to bound their use and reduce the probability of catastrophic misapplication. AI, particularly in lethal autonomous weapons systems (LAWS) applications, demands the same seriousness of engagement.

The core governance challenge is this: AI moves faster than policy. By the time an international treaty is negotiated, ratified, and implemented, the technological reality it was designed to address may have transformed beyond recognition. Domestically, this is also true within the US system of government in various election cycles. This is not an argument against governance. It is an argument for a different approach to governance, one built on principles and red lines rather than specific technical specifications.

There is emerging, if fragile, international consensus around certain baseline principles. Meaningful human control over lethal force decisions. Accountability for AI-enabled actions in conflict. Prohibition on AI systems that target humans based solely on automated identification without human confirmation. These principles appear in US Department of War AI ethics guidelines, in EU AI Act provisions, and in various UN discussions.[11] The challenge is moving from principle to enforceable norm.

The United States has a strong interest in shaping international norms proactively rather than reactively. Nations that wait for others to establish the rules find themselves constrained by frameworks designed to serve other nation's interests. American leadership in AI governance—working through multilateral institutions, bilateral agreements, and industry standards bodies—is not idealism. It is strategic self-interest and flexible realism.

This includes engagement with adversaries. The history of arms control is not a history of trust. It is a history of structured distrust—verification mechanisms, transparency measures, and defined consequences that make cheating costly enough to discourage. The United States and China have begun early-stage dialogues on AI risk. These conversations are preliminary and fragile, but they are necessary. The alternative—two nuclear-armed AI superpowers with no communication channels around AI risk—is far more dangerous than the discomfort of talking to a strategic competitor.

The Societal Dimension: How Democracies Embrace AI Will Determine Whether They Can Compete

National security discussions about AI often focus on governments and militaries. But the question of how societies embrace AI—how citizens understand it, how institutions regulate it, how industries deploy it—is equally determinative of national power.

Democracies face a structural tension that authoritarian states do not. The legitimacy of democratic governance depends on public trust. AI applications that erode privacy, concentrate surveillance power, or produce discriminatory outcomes threaten that trust. [12] The domestic political backlash against AI misuse is not simply a civil liberties concern. It is a strategic vulnerability. A democracy that cannot build public confidence in AI governance cannot sustain the domestic coalition required to fund, develop, and deploy AI at national scale.

This means responsible AI development is not a constraint on national power. It is a condition for it. The United States cannot sustain long-term AI advantage if the domestic workforce distrusts the technology, if allied nations refuse data-sharing partnerships due to governance concerns, or if private sector leaders are reluctant to engage with government due to reputational risk. Trust is infrastructure. The trust begins in public education forums at the earliest opportunity.

The democratic model also offers genuine competitive advantages that are underappreciated in security discussions. Open research ecosystems produce more robust and innovative AI than closed systems optimized for state priorities. Competitive markets drive efficiency and reliability in ways that state-directed development cannot replicate. Allied networks provide access to diverse talent, complementary technology, and shared intelligence. These are not soft advantages. They are hard ones, and they compound exponentially.

Conclusion: Urgency Without Panic

The United States is not losing the AI competition. But it is not winning it decisively either. The gap between American AI potential and American AI performance—in government, in defense integration, in governance architecture—remains wider than the threat environment permits.

What is required is not panic. Panic produces bad policy: overreach, overclassification, isolation from allies and partners, and suppression of the open research ecosystems that sustain American innovation. What is required is urgency, the kind of clear-eyed, sustained institutional commitment that characterized the early Cold War mobilization. The Apollo program was not built by a nation that panicked. It was built by a nation that decided to win.

AI will reshape military power, industrial capacity, intelligence collection, the information environment, and the practice of diplomacy itself. It will do so whether the United States leads that reshaping or not. The question is not whether AI changes the world. It will. The question is whether the principles embedded in that change—about human dignity, accountability, the laws of armed conflict, and democratic governance—reflect American values or someone else's.

That question is still open. The window to answer it is not unlimited.

References

1. Billy Perrigo, "China Is Using AI to Sow Disinformation and Stoke Discord," *Time*, February 2024; and Sara Fischer, "China Has Started Using AI in Influence Operations," *Axios*, September 2023.
2. Brookings Institution, "AI and the Future of Disinformation: Russia's Evolving Influence Operations," 2023.
3. RAND Corporation, *Artificial Intelligence and National Security: The Importance of AI for Decision Advantage* (Santa Monica, CA: RAND Corporation, 2023).
4. Center for Security and Emerging Technology (CSET), *Artificial Intelligence and the Intelligence Community*(Washington, DC: Georgetown University, 2023).
5. Center for Strategic and International Studies (CSIS), *Artificial Intelligence and the War in Ukraine* (Washington, DC: CSIS, 2024).
6. Luis Martinez, "Chinese AI-Enhanced Satellite Imagery Helping Iran Identify Targets in Middle East, U.S. Intelligence Says," *ABC News*, April 2026.
7. Center for Security and Emerging Technology (CSET), *China's 'Intelligentized Warfare' Concept* (Washington, DC: Georgetown University, 2023).
8. Stanford Institute for Human-Centered Artificial Intelligence, *AI Index Report 2024* (Stanford, CA: Stanford University, 2024); and PitchBook, *Global AI Venture Capital Trends* (PitchBook, 2024).
9. U.S. Department of Commerce, *CHIPS and Science Act: Semiconductor Supply Chain Assessment* (Washington, DC: U.S. Department of Commerce, 2024).

10. European Union, *Artificial Intelligence Act* (Brussels: European Union, 2024).

11. National Institute of Standards and Technology (NIST), *AI Risk Management Framework* (Gaithersburg, MD: NIST, 2023).

12. Stanford Institute for Human-Centered Artificial Intelligence, *AI Index Report 2024* (Stanford, CA: Stanford University, 2024).

About the Author

Colonel Gabe Arrington graduated from the United States Air Force Academy with Military Distinction in 2003. He is a senior pilot, having logged over 2,450 hours and 273 combat missions. He has commanded at the Air Task Force and Squadron levels as well as served in multiple positions at the Joint Staff, Air Staff, MAJCOM, Air Operations Center, Wing, Group, and Squadron levels.

Gabe was a key strategist in writing the Joint Warfighting Concept for the Department of Defense spanning multiple administrations. As the Chief of Disruptive Technology for the United States Air Force, he was recognized by the Secretary of the Air Force for leading the #1 team in the Air Force in identifying emerging technologies that led to $3B in funding and a Program Decision Memorandum for an additional $6B establishing a Program of Record. Additionally, Colonel Arrington has been awarded a secondary Air Force Specialty Code in Space Operations after completing the USAF Space Officer Course with the intent of developing multi-domain leaders for the Joint Force.

Colonel Arrington serves on the Board of Directors of Brandeis University, AI 2030, and PPE Kids. He is well known in national security, policy, think tank, and charitable communities for his work across many global issues. He is a Massachusetts Institute of Technology Seminar XXI Fellowship as well as a National Defense Fellow at the Center for a New American Security graduate. He has provided expert commentary and has been published in Defense One, The Hill, Breaking Defense, Real Clear Defense, and Air University.

Email: GabeArrington03@gmail.com

Website: www.Gabe-Arrington.com

CHAPTER 2

FROM MACHINES TO LIVING SYSTEMS: LEADING THROUGH THE AUTOMOTIVE SHIFT

By Saptarshi Basu
Techpreneur championing the Graypreneur economy
Singapore

A machine performs what it is designed to do. A living system learns what it was never explicitly taught. The future of mobility depends on understanding the difference.

—Unknown

A few years back, at a strategy meeting with top car leaders, the talk went as expected. We talked about electrification timelines. We looked at regulatory pressures. We examined competitive positioning. Finally, we explored the drive for better driving automation. The tone was analytical and confident, with a hint of competitiveness.

Then someone posed a question that changed the room's atmosphere: "What happens when the vehicle is no longer finished?" This felt like someone was launching a ballistic missile, and we had no air defence system. Silence fell. This was not a technical question. It was existential.

For over a hundred years, car companies have improved their product development skills. A vehicle moved from design to validation, production, and sale. Once it left the factory, its identity was largely determined. Improvements came in the next model year. The transaction was clear. The ownership experience was predictable.

Now, with artificial intelligence and software-defined architecture, that certainty disappears. The vehicle is no longer static at delivery. You can update, improve, adapt, and even reconfigure it over time. It learns from aggregated fleet data. It changes behaviour based on the environment and usage. It becomes less of a mechanical object and more of a living system.

At first glance, this seems incremental. In reality, it is foundational. It redefines engineering responsibility, reshapes revenue logic, and alters customer trust. It assesses whether an organisation has a clear understanding of systems thinking. Or, is it still stuck on product thinking?

That simple question—*what happens when the vehicle is no longer finished?*—opens the door to the real AI shift.

From Deterministic Machines to Probabilistic Systems

To see why this change feels strange, let's explore what the automotive industry has done well.

For decades, automotive engineering operated within deterministic boundaries. The team locked the specifications. We defined the performance targets in advance. Validation sought to eliminate variance. The culture rewarded certainty. This mindset built one of the most reliable industrial ecosystems in history. Certainty was a strength.

Artificial intelligence introduces a different operating logic. It is probabilistic. It learns patterns from data rather than following static rule sets. Its output quality depends on context, data quality, and continuous refinement. Rather than eliminating uncertainty entirely, AI systems handle it. This difference may seem technical. It is rooted in culture.

For professionals trained to remove ambiguity, this creates tension. Engineers ask if a probabilistic model can be checked. Can it be done like a mechanical subsystem? Compliance leaders question traceability. Executives worry about accountability when decisions are partially automated.

These concerns are legitimate. They are not signs of fear of progress. They are signs of professional integrity. This is why the AI shift in cars isn't about replacing engineers with algorithms. It is about evolving engineering maturity. The question is no longer whether AI belongs in the automotive industry. It clearly does. Will organisations create governance structures? They need to support probabilistic intelligence responsibly. Which brings us to the structural enabler of this transformation.

The Software-Defined Vehicle: The Structural Shift

Public discussions often focus on autonomy. This includes levels of automation, regulatory approvals, and moral dilemmas in extreme situations. But autonomy is only one visible expression of a deeper transformation. The more transformative development is the software-defined vehicle.

In a software-defined architecture, developers abstract core capabilities from hardware constraints. You can activate, deactivate, or enhance features digitally. Over-the-air updates allow continuous improvement. Diagnostics can operate from a distance. Energy management algorithms can evolve based on aggregated usage patterns. The vehicle is not just maintained; it is managed. And when someone manages something, its economics change.

Historically, revenue was concentrated at the point of sale. Optional packages were bundled and paid for upfront. Buyers made decisions based on projected needs rather than lived experience. Often, they paid for capabilities they rarely used.

Software-defined capability allows for a different model. Features can be contextually recommended and temporarily activated. A vehicle could provide better support on highways for long trips. It can also adjust for mountain terrain on weekends. The driver retains choice. Payment aligns with use. The shift may appear operational. It is psychological. Customers no longer need to predict their future perfectly at the time of the purchase. They can adapt. The vehicle grows with them.

For manufacturers, this unlocks recurring lifecycle revenue, but without coercion. It becomes monetisation aligned with intent. The principle is simple: Pay for use, not for promise. Yet, in mobility, this flexibility cannot come at the cost of safety or trust. Which is why architecture alone is insufficient. Governance must evolve alongside it.

Governance as Capability, Not Constraint

The automotive sector is not operating in a regulatory vacuum. International frameworks, such as ISO 26262 for functional safety and ISO/SAE 21434 for cybersecurity, provide clear methods for risk management. UNECE regulations R155 and R156 formalise governance for cybersecurity management systems and software updates. These frameworks exist. Vehicles are changing. They are now digital systems instead of mechanical products. There is a common misconception that governance slows innovation. In reality, governance reduces hesitation.

When update pipelines are secure, teams can move faster. Rollback mechanisms should be tested. Clear roles help everyone know what to do. This keeps the process smooth, not slow. A well-designed governance structure transforms uncertainty into managed exposure. Without governance, innovation stalls. This happens not due to regulators, but because internal stakeholders lose confidence.

Mature AI integration needs clear steps throughout the lifecycle. This includes how data is collected. It explains how models are trained. It shows how bias is checked. It describes how drift is spotted. Finally, it details how accountability is assigned.

Human oversight must be intentional. Human-in-the-loop design is not a temporary bridge until AI becomes flawless. It is a structural design philosophy. Humans define objectives and ethical boundaries. AI scales analysis. Humans interpret outcomes and recalibrate systems. This is not about limiting AI. It is about ensuring trust scales alongside capability. When governance is in place, AI begins to demonstrate its hidden value across the enterprise.

The Governed Intelligence: AI's Real Transformation in Mobility

When people imagine AI in cars, they see a steering wheel moving on its own. They also picture a dashboard lit up with bright graphics. Yet, the real transformation occurring in 2026 is subtler and more profound. Mobility has moved beyond simple automation towards what engineers call "physical AI." These systems don't just handle digital inputs. They also understand, think, and act according to the rules of the physical world.

In the automotive sector, working in the physical world carries significant responsibility. You must treat this with seriousness. It means safety, compliance, and accountability under intense regulatory scrutiny. This is where the story of automotive AI becomes serious.

From Assisted Driving to Contextual Reasoning

Ten years ago, "advanced" systems like adaptive cruise control were rule-based. They were predictable and met safety standards, such as ISO 26262. Today's architectures use deep learning. They also use

Vision-Language Models (VLMs). These tools help interpret complex scenes. Modern vehicles don't just see a pedestrian. They also think about gestures and what a temporary construction barrier means.

Yet, as AI capability increases, regulatory boundaries remain explicit. UNECE rules limit Level 3 (L3) "eyes-off" functionality. It only works in certain highway conditions. It also needs specific weather. This is within its operational design domain (ODD). When someone breaches these boundaries, responsibility must return to the driver without interruption. This is an important governance milestone. Liability transfer and fallback strategies are now real safeguards. They are no longer tech experiments. The machine isn't helping; it's working within clear limits. This shifts the relationship between people and vehicles.

The Cabin as an Intelligent Partner

Inside the vehicle, generative AI has transformed the dashboard into an intelligent partner. Yet, these interfaces do not operate as standalone novelties. They rely on cybersecurity frameworks like ISO/SAE 21434. This helps keep voice inputs and over-the-air updates secure.

When a driver says, "I'm feeling a bit cold," it starts a chain of actions. Climate adjustments and energy optimisation follow, all handled by secure software layers. Integrations with enterprise tools for meetings must follow GDPR. They also need to meet data privacy laws. The cabin is now a safe space that values privacy. Intelligence without governance can be dangerous, but with governance, it becomes empowering.

Interior Intelligence: Monitoring with Boundaries

AI's gaze has expanded from the road to the cabin through driver monitoring systems (DMS). Many safety laws now require these systems to track eye movement and posture. They help detect fatigue or distraction. The aim is not surveillance; it is a safety redundancy. In advanced systems, if the driver does not respond, the vehicle can act.

It will start safe fallback actions. These actions follow clear guidelines for human-machine interaction. Automatic crash notification systems check impact data. They also look at occupancy. This helps emergency services know how serious injuries might be. This data transmission is important. It must follow telecom standards. It must also meet UNECE R155 cybersecurity rules. Even in emergencies, intelligence transmission must be secure. This is what distinguishes industry maturity from mere hype.

Urban Complexity and Fleet Accountability

The urban environment remains the most complex domain for AI. Dense traffic and unpredictable informal gestures make urban autonomy probabilistic and dynamic. Many manufacturers set clear ODD boundaries. They focus on full automation for structured highways. But they stay realistic about city streets.

Fleet operators are leveraging behavioural AI more than ever across their entire fleets. Telematics systems measure braking intensity and cornering to refine risk models for insurers. This aligns with usage-based insurance logic, but its responsible deployment requires transparency. Drivers need to know how data impacts premiums. Policies should explain anonymisation in a straightforward manner. This model works because AI helps measure accountability. It cuts down on systemic risk and boosts operational discipline.

The Quiet Revolution

A pattern is now visible across the industry. AI in cars has changed. It now goes beyond basic automation. It offers smart, governed intelligence. This includes things like contextual autonomy, smart cabins, and predictive maintenance. It listens and interprets. It also anticipates. But it always stays within clear safety, cybersecurity, and operational limits.

Most drivers will never read an ISO standard or study a UNECE annex. They will see that their vehicle feels more aware. It feels more supportive, too. Plus, it's less surprising. AI's true value lies not in showmanship, but in stability. It doesn't aim to replace humans; instead, it supports the systems that surround them. When intelligence operates within disciplined governance, trust increases consistently alongside it. The vehicle has shifted from reckless to in control. That's where we create real value.

The Buyer Has Already Changed

Big changes are happening in the showroom. This goes beyond boardroom talks about standards and design. In 2026, the market shows a "great split" in consumer readiness. Buyers welcome AI that makes life easier, but are wary of AI that takes away their control.

The Three Tiers of Consumer Readiness

Buyer acceptance happens in stages. It depends on risk levels and the extent to which human authority is retained.

- *High Readiness: The Digital Assistant:* Nearly 90% of buyers want AI voice systems in their cars. Trust is strong here since the stakes are low. If an AI picks a bad café or misreads an email, the driver can quickly fix it.
- *Moderate Readiness: Supervised Automation:* About 40% to 45% of buyers want Level 2+ for highways. Many prefer this level for added safety and safer driving. Consumers view these as tools for reducing fatigue, not just gadgets. However, they need to stay "eyes on" and keep ultimate responsibility.
- *Low Readiness: The Trust Wall*: Around 25% of buyers feel okay with "eyes-off" (Level 3) features. Most are not. Even with approvals, the "black box" problem is an issue. Drivers feel unsure. They hesitate to let a system make life-and-death choices. They don't understand how it works.

Emerging Barriers to Adoption

Aside from the technology, three key issues are influencing the 2026 buyer's journey:

1. *Subscription Fatigue:* One in three buyers is resistant to "pay-to-play" models. They are willing to pay for clear value but resist paying for access to hardware they already own.
2. *Data Sovereignty:* Vehicles track biometric data such as heart rate and eye movement. The question "Who owns my data?" has become a deal-breaker. Transparency and consent are now prerequisites for trust.
3. *The Gimmick Filter:* Buyers are tired of flashy AI, like augmented reality. They now look at features in a practical way: Does it make driving safer, easier, or more reliable?

Safety Over Sparkle

The 2026 buyer is not seeking a spaceship; they are seeking a safety shield. They like AI that stops collisions. It also helps reduce fatigue. But they are cautious about systems that add financial friction. They do not trust those that are not transparent. For manufacturers, success in 2026 isn't about having the most advanced AI. It's about having the most trusted AI. This trust is built on three pillars: control, transparency, and utility.

The Deeper Shift

The automotive industry is shifting from a product mindset to a stewardship mindset. This change means moving from frozen validation to lifecycle accountability. The true measure of maturity in 2026 is not the advancement of the AI itself. Instead, it is the safety, transparency, and ethics with which it operates.

Thriving in the Shift

This era's winners will define themselves primarily by the speed of their automation. They will be the organisations that design systems. These systems will show strong responsibility. They will also build trust into their structures. AI acts as an amplifier. It speeds up organisations with clear governance and integration. But it makes those in siloed structures hesitate.

The Industry Crossroads

The automotive sector currently stands at a critical juncture between two distinct paths:

- *The Feature Mindset:* Seeing AI as a checkbox, an upgrade badge, or a slide in an investor's deck.
- *The Ecosystem Mindset:* View mobility as a digital ecosystem. It needs long-term care, humility, and maturity.

Thriving requires embracing adaptability as a core discipline rather than a marketing buzzword. Vehicles should upgrade with minimal noise using software. Monetisation must match the value given. We should manage data with full transparency.

From Control to Stewardship

The biggest change is the shift from control to stewardship in leadership. The industry has spent years perfecting mechanical precision in steel. Now, it must also handle new, learning, and interactive systems.

- *Long-Term Relationships:* Replacing one-time transactions with ongoing partnerships.
- *Adaptive Ownership:* Moving from static products to systems that grow with the user.
- *Governance Over Fear:* Replacing the uncertainty of AI with confidence in disciplined oversight.

The Real AI Opportunity

The future will not define the vehicle only by its autonomy. Instead, its trust, adaptability, and shared value will define it. This transition is moving the industry from selling machines to stewarding intelligent systems. The real change isn't about technology. It's about ethics, organisation, and relationships. It also covers how buyers and their ecosystem accept these ideas. The question isn't if this change will happen. It's about whether industry leaders will provide smart, responsible advice.

Acknowledgement Note

This work isn't chapters. It's a guide from a 30-year journey in the "infinite classroom" of our industry. These words are collective. The wisdom comes from many mentors, colleagues, and experts I have worked with.

These ideas—from complex probabilistic systems to the role of ethics in AI—didn't appear out of thin air. The global OEMs shaped them. They were fine-tuned in agile startups. Researchers discussed them in quiet labs.

I owe a huge thanks to my "Gray Hair" mentors. They shared their insights when I was starting, and I was just another face in the crowd, a "nobody" with an engineering and management degree and a lot of questions. I am also grateful to my brilliant peers. They supported me during big multimillion-pound changes. Every "battle scar" from a failed system and every success from a launch have added to this pool of knowledge.

Thank you to all who shared their skills. Your knowledge and insights are of immense value. This book celebrates our shared intelligence. It reminds us that technology moves fast. But it's the human connection that keeps it safe and meaningful.

About the Author

Saptarshi Basu is a tech entrepreneur. He is also an automotive strategist. For nearly 30 years, he has worked in global OEMs, fintech, and AI-driven ecosystems. He has led multimillion-dollar programs in software-defined mobility and connected commerce. Still, his main focus is on the people behind the machines.

Saptarshi is a passionate advocate for the "Graypreneur" movement. He focuses on more than the technical aspects of engineering. He defends the "Wisdom of the Grey Hair" with great fervour. He questions the industry's trend. It often ignores experienced professionals because of budget cuts. To Saptarshi, this isn't a corporate shift; it is the creation of a dangerous "knowledge chasm." He thinks that pushing aside experienced talent breaks the chain of knowledge transfer. This makes companies skilled but lacking in judgment. They miss the valuable memory that years of experience provide.

Saptarshi is the founder of a fintech start-up and OrbitNext Consulting. He works at the intersection of AI governance, Fintech, and automotive innovation. He advises organisations on building smart systems. These systems lack adaptability; they also undergo responsible management. He promotes a workforce that sees the "longevity dividend." This means valuing senior experience as a valuable strategic asset, not a cost to reduce.

Saptarshi enjoys talking about work, wisdom, and technology. He is always open to these conversations. You can reach him at saptarshi@basu.global directly.

CHAPTER 3

EVERYONE BECOMES A BUILDER

By Nati Berkover
AI Builder, Entrepreneur, and Product Leader
Tel Aviv, Israel

The future is already here—it's just not evenly distributed.
—William Gibson

The New Job Description

In May 2025, Dario Amodei, the CEO of Anthropic (the company behind Claude, one of the most advanced AI systems in the world), gave an interview to Axios and said something that should keep every professional up at night: AI could eliminate half of all entry-level white-collar jobs within five years. Not blue-collar jobs. Not factory workers. *White-collar* jobs. The kind of jobs that require a degree, a suit, and a computer.

He wasn't alone. A few months later, Mustafa Suleyman, Microsoft's AI chief, predicted that virtually all professional tasks performed at a computer would be automated within 18 months. Ford's CEO Jim Farley said AI would cut white-collar headcount in half. These aren't bloggers or futurists. These are the CEOs of the companies actually building this technology.

Now, if you're reading this and thinking, "Okay, but people have been saying this about every new technology for decades," I hear you. But let me share something that changed how I think about this.

A 2025 Thomson Reuters report looked at lawyers, accountants, and auditors who were already experimenting with AI tools. The result? Marginal productivity improvements at best. In some cases, AI actually made workers *less* productive. A separate study by the Model Evaluation and Threat Research (METR) institute found that software developers using AI coding assistants took 20% longer to finish their tasks. Twenty percent *longer.*

So is AI overhyped? No. What these studies reveal is something much more important: Simply bolting AI onto your existing workflow doesn't work. Using ChatGPT to write your emails a bit faster or asking Copilot to autocomplete your code is not the transformation. That's like buying a smartphone in 2008 and only using it to make phone calls.

The people who are actually thriving right now aren't the ones who learned to "use AI tools." They're the ones who realized that their job is no longer to do the work but to build the systems that do the work.

I call this concept the "specialized builder." The idea is simple: In the age of AI, every professional becomes someone who builds solutions within their domain of expertise. Your domain knowledge doesn't become less valuable. Quite the opposite. It becomes the blueprint for what you build. And if you don't make this shift, someone else in your field will, and they'll do your job and theirs.

From Fear to Building

Let me tell you how I know this works, not from a TED talk or a research paper, but from doing it myself.

As of the time of writing this chapter, I manage the data group at my company. It's made up of three teams: data analytics, data engineering, and data science. Without getting into technicalities, these are the people responsible for everything from building dashboards and reports, through managing the company's data infrastructure, to developing the AI and machine learning models that power our products.

At our recent team offsite, I delivered a message that made some people uncomfortable: Each of these teams needs to transform how they work. Not *what* they work on their core mission stays the same. But *how* they deliver on that mission needs to change fundamentally.

For the analytics team, it meant moving from building dashboards people have to check, to building tools that *push* insights to stakeholders proactively. For the engineering team, it meant moving from making data available to making data *usable*, specifically for AI systems and applications to consume. For the data science team, it meant moving from developing models in isolation to owning end-to-end AI systems that others can configure and interact with.

In all three cases, the direction was the same: Stop being someone who executes tasks. Start being someone who builds the systems that handle those tasks. The whole company benefits when each of these teams levels up because their work feeds directly into the products and decisions that drive the business.

Now here's the part that might surprise you. You'd think that people working in tech, surrounded by AI every day, would be the first to embrace this shift. But the reality was the opposite. Many of my team members were genuinely afraid. And in some ways, their fear made perfect sense. If you work in technology, it's easy to look at AI and think, "This thing can already do significant parts of my job. Am I next?"

If even tech professionals who work with data and AI daily are afraid, imagine how the rest of the workforce feels. The fear is universal. And it's the first thing you have to address before any transformation can happen.

Getting Your Hands Dirty

So how do you take a group of smart, capable, but fearful professionals and bring them to the other side of this transformation?

You don't start with a PowerPoint about the future of AI. You don't send them to a training course. You make them build something.

The first step was an internal exercise we designed called "InsightFlow." It was deliberately the same assignment for everyone across all three teams: Build a complete application, end-to-end, that pulls real company data, applies AI-powered analysis, and delivers a useful insights report. Not a script. Not a prototype for a meeting. A real, working application with a user interface that someone else could actually use.

The point of making it identical for everyone wasn't to test technical skill. It was to establish a baseline. I needed every single team member to be able to say, with confidence: "I built an application with my own hands, from scratch, and it works." That's it. That was the first goal. Cross the fear barrier. Prove to yourself that you can do this.

And they did. Within weeks, people who had never built a web application in their life were deploying functional tools. Not because they suddenly became software engineers, but because AI itself was their development partner. They used AI coding assistants to write code, debug issues, and learn new frameworks on the fly. The irony was beautiful: They were using AI to learn how to build with AI.

But that was just the beginning. The InsightFlow exercise was the warm-up. The real transformation came next.

As part of the yearly goals and our group's objectives, each team had to build several fully functional applications or services that would help either internal or external stakeholders do their work better. Not experiments. Not proofs of concept. Real tools that real people would depend on.

This is where the shift happened. When your analytics team builds a tool that automatically surfaces revenue anomalies and sends a Slack message to the VP of sales before they even know something changed, that's not a dashboard. That's a product. When your data engineering team builds a self-service data catalog that lets product managers find and query data without filing a ticket, that's not a pipeline. That's a product. When your data science team builds an application that lets marketing configure and run audience segmentation models through a simple interface, that's not a model. That's a product.

And when all of that is happening across your organization? Your people can finally, rightfully, call themselves builders.

The Pattern

Looking back at this journey, I can see a clear pattern that I believe applies to any team, in any industry. Here's how the transformation actually unfolds:

It starts with fear. Fear of change. Fear of the unknown. Fear of losing your job or becoming irrelevant. This is real and it's universal. Don't dismiss it and don't pretend it doesn't exist. Acknowledge it head-on.

Then comes acknowledgment. A revolution is taking place. It's not slowing down. And shoving your head in the sand is probably the worst thing you can do. As a leader, your job is to make this undeniably clear, not to scare people, but to create the urgency that makes the next steps feel necessary rather than optional.

Next, you reframe the narrative. This is the most important step. You don't say, "Everything is changing and you need to keep up." You say, "Your mission stays the same. The way you accomplish it is about to get significantly better." Fear of replacement turns into excitement about amplification.

Then you show them how, by getting your own hands dirty first. This is critical. You cannot ask your team to become builders if you haven't built anything yourself. When I asked my teams to build applications, I had already done it. I could speak from experience, troubleshoot problems, and most importantly, I could show them it's possible because I'd done it myself. Lead by example. Always.

After that, you hold their hands through the transformation. Start with a simple, identical exercise for everyone to establish a baseline. Make it safe to struggle. Make it safe to ask for help. The goal isn't to produce perfect software. It's to break through the fear barrier.

Then you raise the stakes. Move from exercises to real projects with real stakeholders. Set clear objectives. Make building part of the team's formal goals, not a side project or a "nice to have."

And finally, you celebrate. Celebrate the tools and applications that were built. Celebrate the new impact delivered. Focus on the outcomes: This is what we built, this is who it helps, this is the value it creates. And call your people what they now are: builders.

That's the playbook. Fear, acknowledgment, reframing, leading by example, supporting the transition, raising the bar, and celebrating the results. It worked for my tech teams. But I'm convinced it works far beyond tech.

It's Not Just for Tech Teams

You might be reading this and thinking, "That's a nice story, but I don't run a data department. I'm not in tech. This doesn't apply to me."

I'd argue it does. Let me show you.

The bookkeeper. Their mission has always been the same: Maintain accurate, up-to-date financial records. Before AI, that meant manually categorizing transactions, reconciling bank statements, chasing down receipts, and preparing reports. The bookkeeper of tomorrow doesn't stop doing these things, but they stop doing them manually.

Instead, they become what I'd call a "financial data quality builder." They configure AI-powered categorization rules. They build exception-handling workflows that flag anomalies for human review.

They design automated reconciliation systems that run continuously rather than once at month-end. The bookkeeper who builds a system that catches errors in real time and produces weekly financial health summaries without being asked is not doing less valuable work. They're doing radically *more* valuable work, and they're doing it at a scale that wasn't possible before.

The procurement manager. Their mission: Source the right goods and services at the best price from reliable suppliers. Traditionally, this meant managing spreadsheets of vendors, manually comparing quotes, tracking contract renewals on calendar reminders, and building relationships through phone calls and emails.

The procurement manager of tomorrow becomes a "supplier intelligence builder." They design automated vendor evaluation tools that score suppliers on delivery time, quality metrics, and pricing trends. They build spend analysis dashboards powered by AI that surface savings opportunities nobody asked about. They create alert systems that notify them when a contract is approaching renewal or when market conditions suggest renegotiation.

Same mission. Same domain expertise. Completely different leverage.

A True Builder Mindset Has No Boundaries

And if you believe these professional examples are not applicable to your field or business, that's completely fine. A true builder mindset doesn't stop at the office door. Once you start thinking this way, it expands into your personal life as well.

Here's a quick example. I recently renovated my apartment. I asked three contractors to submit offers for the project. As you can imagine, even though I did my best to provide the requirements in the most detailed and structured way possible, each contractor used their own terminology and priced similar items completely differently. Comparing their offers was like comparing apples to oranges to kiwis.

What did I do?

AI to the rescue.

Having a builder mindset, I asked an AI assistant to build me a utility and a dashboard that compared the three offers side by side, in a way that made the comparison apples-to-apples. I uploaded the offers, answered a few clarifying questions, and provided some research I'd gathered on standard renovation pricing. Within 15 minutes I had an interactive dashboard that allowed me to compare the offers, upload additional quotes, and export everything to a spreadsheet.

Fifteen minutes. No coding skills required. I didn't ask AI to "summarize" the offers or "recommend" a contractor. I asked it to *build me a tool* that helps me make that decision myself, with better data. That's the builder mindset. And it works whether you're managing a data team, running a procurement department, or renovating your apartment.

And here's the part that surprises most people: You don't need to become a software engineer to think this way. The same AI that's transforming your job can help you build the tools that transform it. The barrier is not coding ability. The barrier is the decision to start building.

The AI revolution is not something that's happening to us. It's something we get to shape, if we choose to.

The professionals and leaders who will thrive in the next decade are not the ones who become expert *users* of AI. They're the ones who become *builders* with AI. Your domain expertise is your superpower. Your mission hasn't changed. But your job description has, whether you rewrite it yourself or someone else rewrites it for you.

So, are you ready to become a builder?

About the Author

Nati Berkover is an entrepreneur, product leader, and AI builder with over 25 years of experience in the high-tech industry. Starting as a software engineer specializing in algorithms and large-scale systems, he later transitioned into product leadership, where he led product, data, and AI initiatives at scale.

Nati is the co-founder of Newsfusion, an AI-driven news aggregation platform acquired in 2020, and has held executive roles including VP of product and data, leading multidisciplinary teams across product, data science, and engineering.

Today, Nati advises companies on product strategy and AI transformation, helping them rethink how work gets done in the age of AI. His focus is on turning professionals into builders, people who design and create systems that replace manual work rather than perform it.

In parallel, he builds AI-powered products himself, using modern AI tools to dramatically accelerate development and challenge traditional assumptions about how software is created. His work focuses on the intersection of product, data, and AI, with a strong emphasis on execution and real-world impact.

Email: nati@pmeffective.com

Website: https://effectiveproductmanagement.com/knowledge-base-get-smarter/

LinkedIn: https://www.linkedin.com/in/natib/

CHAPTER 4

THE AI INDUSTRIAL RESET: BUILDING FACTORIES OF INTELLIGENCE

By Phani Burra
VP, AI & Digital; Architecting AI-First Enterprises
Dublin, California

Civilization advances by extending the number of important operations which we can perform without thinking about them.
—Alfred North Whitehead

From Invention to Infrastructure

Every great technological era begins with excitement and matures with construction. The first wave of artificial intelligence dazzled us with possibility. Models wrote text, generated images, and answered questions. It felt light, fast, almost magical. The barrier to entry appeared low, and the outputs felt immediate. For many, it seemed as if intelligence itself had suddenly become abundant and accessible. That phase is ending.

As AI becomes mission-critical, civilization must do what it has always done next: Build infrastructure. The shift from experimentation to reliability is not optional. It is inevitable. When technologies begin to underpin economic systems, national competitiveness, and daily operations, they must be engineered for consistency, durability, and trust. This is the AI industrial reset.

In earlier eras, invention captured imagination, but infrastructure created impact. The steam engine mattered, but railways transformed nations. Electricity was revolutionary, but power grids made it indispensable. Likewise, AI models may inspire, but AI infrastructure will define outcomes.

What once felt experimental must now become dependable. The transition from novelty to necessity marks the moment when technology stops being a curiosity and starts becoming embedded into the fabric of daily life, business, and governance. The organizations and societies that recognize this shift early will build advantages that compound over time.

Intelligence as a Means of Production

AI is no longer an experiment. It is becoming a means of production. Just as factories transformed raw materials into physical goods, modern AI systems transform raw data into decisions, predictions, and automated action. Intelligence is manufactured continuously, not generated once. It flows through systems, improves through feedback, and scales through replication.

This represents a profound shift. In previous eras, intelligence was primarily human and constrained by time, attention, and expertise. Now, it can be codified, distributed, and amplified. Organizations are no longer limited by how many experts they can hire, but by how effectively they can operationalize intelligence across their systems.

This requires reliability, scale, governance, and energy, traits of industry, not software demos.

The language must change. AI is not just a tool; it is a production system. Outputs must be measured. Performance must be monitored. Failures must be anticipated and mitigated. Continuous improvement must be engineered into the system itself.

Organizations must now think less like innovators chasing breakthroughs and more like operators ensuring consistent output, uptime, and measurable impact. The question is no longer "What can AI do?" but "How do we run AI reliably at scale?" Those who answer this question effectively will not only increase efficiency; they will redefine what productivity means in the modern economy.

The Hidden Backbone

Behind every AI success story lies invisible labor:

- Compute infrastructure
- Networks and interconnects
- Energy and cooling systems
- Governance and security frameworks
- Human oversight and operational discipline

These are not optional components. They are load-bearing.

The popular narrative around AI often focuses on models and interfaces, but the real work happens beneath the surface. Data pipelines must be built and maintained. Systems must be monitored in real time. Failures must be detected and corrected before they cascade. The organizations that master this hidden layer will lead. The ones that ignore it will depend.

This backbone determines not only performance, but also trust, resilience, and long-term viability in an increasingly intelligence-driven world. It is the difference between a system that works in a demo and one that operates reliably in production. Just as factories required supply chains, maintenance crews, and quality control processes, AI systems require continuous operational support. There is no "set it and forget it" in an intelligence economy. There is only sustained effort, disciplined execution, and ongoing refinement. The most successful organizations will treat this backbone as a strategic asset, not a technical afterthought.

Power Becomes Destiny

Every industrial era has its bottleneck. For AI, it is energy. Training and running intelligent systems consumes enormous power. As demand grows, access to reliable energy becomes a strategic differentiator. Compute capacity and energy availability are increasingly intertwined, forming the foundation upon which intelligence is produced.

This reshapes geography, economics, and geopolitics. Regions with abundant energy resources, advanced data infrastructure, and supportive regulatory environments will attract investment and innovation. Others may find themselves constrained, not by lack of talent, but by lack of capacity. Infrastructure is no longer background. It is destiny.

In this new landscape, decisions about energy policy, data center placement, and resource allocation carry long-term strategic consequences. The competition for energy-efficient computing, renewable power integration, and optimized cooling systems will intensify.

Nations and organizations that secure energy and compute capacity will shape the balance of power in this new era. Those that fail to do so may find themselves dependent on external providers for the very intelligence that drives their economies.

The industrial reset is not just technological; it is structural. And at its core lies the question of who can sustain intelligence at scale.

Governance Is Not Bureaucracy

In an intelligence economy, governance is infrastructure. Without trust, intelligence cannot scale. Without transparency, automation becomes liability. Governance systems—monitoring, controls, accountability—are as essential as servers and networks.

Responsible AI is not an ethical add-on. It is an operational requirement. As AI systems make decisions that affect people, organizations, and societies, the need for clear accountability becomes paramount. Who is responsible when a system fails? How are decisions explained? How are risks identified and mitigated? Clarity in decision-making, auditability of systems, and alignment with human values are what allow intelligence to be deployed confidently at scale.

Governance enables adoption. It creates the conditions under which organizations can trust their systems, regulators can provide oversight, and users can engage without fear. In this sense, governance is not a constraint. It is an enabler. It transforms AI from a source of uncertainty into a foundation for growth.

The Workforce Reset

Industrial shifts do not eliminate work. They transform it. The AI era creates demand for builders, operators, and stewards of intelligence:

- Platform engineers
- AI operations leaders
- Security and risk professionals
- Domain experts who translate intelligence into outcomes

Resilience comes from systems thinking, not tool mastery. The workforce of the AI era will be defined not by narrow specialization, but by the ability to integrate across disciplines. Technical expertise must be combined with domain knowledge, operational awareness, and strategic thinking.

The most valuable contributors will be those who understand how to connect technology, people, and processes into cohesive, functioning systems. They will not simply use AI. They will shape how it is deployed, governed, and improved.

At the same time, organizations must invest in reskilling and education. The transition to an intelligence-driven economy requires new capabilities, new mindsets, and new forms of collaboration. The workforce reset is not about replacement. It is about evolution.

Closing Reflection

The most important work of this era is not glamorous. It is the quiet, disciplined effort of building foundations strong enough to carry intelligence at scale. It is the work of engineers who design resilient systems, operators who ensure uptime, leaders who create alignment, and teams who continuously improve what has been built. The future will belong not just to those who invent intelligence, but to those who build the factories that sustain it.

About the Author

Phani Burra is an AI and digital transformation leader with over two decades of experience driving enterprise-scale modernization. With a master's degree in computer science and advanced leadership training from Harvard Business School, he combines deep technical expertise with strategic executive insight. His work focuses on generative AI, platform engineering, DevSecOps, and automation-led transformation to help organizations unlock measurable business value.

Phani actively engages with leading academic and executive ecosystems, including contributions to Harvard Business Review discussions and executive learning forums, and participation in programs associated with MIT Sloan. He brings a practitioner's lens to AI strategy, enterprise transformation, and board-level technology alignment.

Known for building high-performing global teams and managing complex transformation portfolios, Phani is passionate about responsible AI adoption, innovation at scale, and shaping the future of intelligent, resilient enterprises.

Email: phani.burra@gmail.com

LinkedIn: https://www.linkedin.com/in/phaniburra/

CHAPTER 5

ARCHITECTING TRUST AND ENGINEERING TRUTH

By Chris Clarkson
AI Technologist and Consultant
Adelaide, Australia

The saddest aspect of life right now is that science gathers knowledge faster than society gathers wisdom.

—Isaac Asimov

AI is not a single technology; it is a set of demands that land squarely on the physical and organisational infrastructure of modern organisations. The headlines focus on models, breakthroughs, and applications. The harder, less visible work is deciding whether those ambitions can be sustained by power, cooling, procurement cycles, and system and data governance and then building the systems and relationships to turn them into reality. This chapter is written for leaders who must translate excitement into durable decisions: CIOs, infrastructure owners, procurement teams, data owners and technical leaders who want to turn AI opportunity into long-term value without creating structural debt.

The Data Reality: Garbage In, Garbage Out

Too often, AI conversations start with the software stack and end there and that's a mistake! Compute-intensive workloads associated with AI training, large-scale inference or quantum experimentation place sustained demands on power, cooling, procurement cycles, and vendor dependency.

These constraints are not footnotes; they are the governing variables of success. When designing a roadmap for advanced computing, you are designing a capital program that will outlive product cycles and executive tenures. Poor decisions such as underpowered facilities, a procurement cadence that lags GPU roadmaps, or a vendor lock-in that prevents migration are expensive and often irreversible.

This is where "engineering truth" matters. Engineering truth is the honest assessment of what a system will deliver under load, over time, and at scale. It is the discipline of matching ambition to reality before contracts are signed and press releases are written.

Technical Empathy as a Leadership Framework

The leadership posture required here, what I call "technical empathy," is the ability to hold engineering constraints and commercial objectives in the same room and translate between them without distortion. At its core, technical empathy means that commercial ambition must never get ahead of engineering reality.

Technical empathy has four practical pillars:

- **Data Integrity**: The organisation has data sets to reflect its needs and perspectives into the training models, rather than relying on external data to solve its problems and questions.

- **Engineering Integrity:** The organisation validates that the architecture can sustain the workload. If the design cannot hold under expected demand, no amount of commercial optimism will save the project.
- **Bidirectional Translation:** The organisation speaks both languages fluently. It translates technical risk into commercial terms and commercial goals into technical requirements enabling informed trade-offs across stakeholders.
- **Failure-Mode Forecasting:** The organisation maps downstream consequences of shortcuts. It shows the outages, cost overruns, and reputational damage that follow when corners are cut.

This is not a framework for slide decks. It is the way viable decisions are made in rooms where budgets, regulations, and engineering realities align.

Three Filters to Separate Signal from Noise

Every vendor pitch claims to be transformative. Most are not. Separate genuine structural shifts from expensive distractions by applying three filters before committing capital:

1. **Physics:** Does the proposition respect the immutable constraints of thermals, latency, and power economics? If it violates basic physics, it is a fad regardless of the marketing.
2. **Procurement Cycles:** Can your organisation build a realistic five-to-ten-year budget around this technology? If procurement cannot accommodate the cadence of upgrades, you will be perpetually behind.
3. **Ecosystem Commitment:** Are hyperscalers, chipmakers, and regulators moving in the same direction? Structural shifts require broad ecosystem alignment to be durable.

Use these filters early and often. They are not academic; they are the difference between a resilient platform and a stranded asset.

Designing for Readiness Not Hype

A readiness-led approach begins with people, not vendors. It begins by interrogating real demand: end-user interviews, workload profiling, and a clear view of how researchers, data scientists, and production teams will consume compute over time. Training bursts, steady inference, latency-sensitive edge workloads, and archival needs must all be explicitly understood. From that evidence, a mix of CPU, GPU, and emerging resources can be designed.

Design choices should be explicit about trade-offs:

- **Hyperscale for training and heavy batch workloads:** Concentrated power, economies of scale, and efficient cooling make hyperscale facilities the right home for large model training.
- **Edge for latency-sensitive inference:** When milliseconds matter, distributed edge sites are the reflex system that keeps applications responsive.
- **Hybrid for resilience and cost control:** By combining centralised training with distributed inference, you match workload characteristics to the right infrastructure.

Economics and failure modes must be modelled, not assumed. Stakeholders need to see the impact of delayed GPU generations, rising power costs, or the loss of a product line from a critical vendor. When the trade-offs are visible, sensible decisions follow.

Putting a Number on the Invisible (But Inevitable)

Professional services and advisory work are often treated as optional line items. That view is short-sighted. Expertise reduces risk, accelerates delivery, and transfers capability. Investment in advisory and design should be justified using three concrete measures:

- **Risk Avoidance Metrics:** Quantify the cost of outages, rework, and misprovisioned capacity.
- **Acceleration Metrics:** Measure how expert-led projects shorten time-to-value.
- **Capability Transfer:** Demonstrate how advisory work builds internal skills that compound over time.

When these measures are presented alongside hardware and software costs, advisory services stop being discretionary and become a multiplier on the investment.

Governance, Accountability, and the Moral Dimension

AI systems will make consequential decisions. That reality demands governance that is as operational as it is ethical. Governance must answer three questions before deployment is permitted:

1. **Who is accountable?** Define clear ownership for outcomes, not just for components.
2. **What are the failure modes?** Enumerate scenarios and assign responsibilities for detection and remediation.
3. **How will decisions be audited?** Ensure traceability for model inputs, training data, and decision logic.

Accountability cannot be outsourced to a model. It must be embedded in contracts, operational runbooks, and executive dashboards. That is how organisations avoid moral and legal ambiguity when systems fail.

Practical Steps for Leaders

Leaders responsible for turning AI ambition into reality will find the following checklist a pragmatic guide to execution:

- **Run an end-user readiness audit including the state of your data.** Interview the people who will use compute daily. Capture workload profiles and growth assumptions.
- **Map infrastructure constraints.** Quantify power, cooling, and space requirements for the next three GPU generations you expect to deploy.
- **Align procurement to roadmaps.** Build procurement cycles that match vendor release cadences and avoid being locked into obsolete hardware.
- **Model hybrid architectures.** Design for hyperscale training and edge inference where appropriate; avoid binary decisions.
- **Create a failure-mode register.** List the top ten ways the project can fail and assign owners and mitigations for each.
- **Invest in translation capability.** Ensure someone in the organisation can translate technical risk into commercial terms and vice versa.
- **Measure advisory ROI.** Track risk avoidance, acceleration, and capability transfer to justify expert involvement.

These are not theoretical exercises. They are the operational steps that keep projects advancing and budgets honest.

The Long View

Technology cycles will continue to accelerate. GPU roadmaps will shift, new accelerators will appear, and quantum experimentation will mature. The organisations that thrive will be those that treat compute as national infrastructure, strategic, durable, and governed, rather than as a series of tactical purchases.

This requires a cultural shift: from chasing the fastest processor to building systems that endure. It requires leaders who can be both innovation-hungry and hype-resistant. It requires the discipline to ask hard questions early and the humility to listen to engineers when they say a design will not hold.

Conclusion

AI offers extraordinary potential, but potential without readiness is risk. The job of leadership is to convert possibility into sustainable capability. That conversion happens when commercial ambition is held in balance with engineering reality, when decisions are grounded in end-user evidence, and when governance and accountability are built into the fabric of the program.

If you take one thing from this chapter, let it be this: Design for readiness, not for headlines. Build the infrastructure, the processes, the data sets and the relationships that let AI deliver value year after year, not just for a press cycle.

References

Selected lines from source material used in this chapter:

- "Technical empathy, to me, means that commercial ambition must never get ahead of engineering reality."
- "Compute-intensive workloads tied to AI training, large-scale inference or quantum experimentation place sustained demands on power, cooling, procurement cycles and vendor dependency."

About the Author

Chris Clarkson is principal consultant and founder of double c consultants. With over 35 years of experience in high performance computing and infrastructure strategy, he helps organisations translate advanced computing ambition into durable, infrastructure-ready decisions across the Asia-Pacific region. Chris specialises in aligning technical realities with commercial objectives, advising on GPU roadmaps, data centre readiness, and hybrid compute architectures. He is known for a pragmatic, evidence-led approach that prioritises long-term resilience over short-term hype.

Email: chris@cc-consultants.com.au

Website: www.cc-consultants.com.au

CHAPTER 6

HUMAN JUDGEMENT, AI POWER, AND THE RACE FOR OPERATIONAL EXCELLENCE

By Alex Dean
AI Advisor and CIO Whisperer
Zurich, Switzerland

Nothing is so useless as doing efficiently that which should not be done at all.

—Peter Drucker

There is a truth that no one in the technology vendor ecosystem will put in a brochure: AI does not transform organisations. People do. And people only change when they understand why the change matters, not just what the new tools do. The organisations that have succeeded with AI transformation started with a belief about what they wanted to become. The technology was the vehicle. The belief was the engine.

I have been having versions of this conversation for over 20 years, first as a technology practitioner, then as someone who sat with CIOs, CDOs, and heads of AI trying to figure out why transformations that looked right on paper kept dying on the factory floor. The pattern is consistent. Deloitte surveyed more than 3,000 senior leaders and found that only 34% of organisations are genuinely reimagining how they work with AI. The rest are layering new tools onto old structures. They are taking a slow process and making it slightly faster. Digitising the paper trail instead of asking whether the paper trail should exist at all. That is not transformation. They are trying to create faster horses.

For manufacturing companies, this gap is particularly costly. The Japanese understood decades ago that competitive advantage is not built on technology but on a mindset: that the person closest to the problem is best placed to solve it, and that continuous improvement is everyone's responsibility. That belief produced Kaizen, Toyota Production System, and a quality revolution that Western manufacturers are still catching up to. Not because they lacked the tools, but because they never fully absorbed the thinking behind them.

AI is the next version of that test. And most European manufacturers are failing it for exactly the same reason they struggled with Industry 4.0: treating a mindset challenge as a technology procurement decision.

This chapter is a transformation roadmap built around four shifts: (1) getting the board to own the ambition; (2) rewiring how decisions are made; (3) restructuring around cross-functional pods where forward deployed AI engineers work alongside subject matter experts; and (4) planning honestly for a workforce that includes agents as a genuine resource. Each shift is hard. None is optional.

Part One: The Board Has to Own This

The most dangerous sentence in corporate AI strategy is "Let's give this to the CIO."

Not because CIOs are incapable. On the contrary, most are excellent at designing and implementing strategies. It is dangerous because of what the sentence reveals: The board believes this is a technology problem. The moment that framing takes hold, the transformation is already compromised. AI transformation is not a technology problem. It is a question about what kind of organisation you want to be and whether leadership has the conviction to pursue it.

Philip Morris International understood this. Their AI@PMI programme did not sit in IT. Every function developed its own strategy, aligned to the company's transformation goals, with genuine shared ownership across technology, strategy, and people and culture. Forrester gave them the Enterprise Architecture Award for EMEA in 2025. Not because the technology was impressive. Because the governance was real. The transformation belonged to everyone.

Galenica told a similar story. When their journey was shared publicly, it was not the CIO who spoke. It was their chief sustainability and transformation officer, describing how they had physically and structurally pulled out the rugs at the top floor, building a collaborative leadership mindset before asking anyone below to change how they operated. That detail matters. It tells you where transformation actually lives: not in the IT budget, but in the room where the executive team decides what kind of company they want to be.

If that conversation is not genuinely on the agenda of every leader in that room, not delegated, not parked in a working group, then what you are building is not a transformation. It is a very expensive experiment that your CIO will eventually be blamed for.

The sequence that works: the board sets the ambition, the executive team owns the execution, the CIO enables the capability. In that order. Reverse it and you get pilots without direction, technology without strategy, and a growing library of proof-of-concept projects that proved nothing except the organisation was not ready to act.

The strategic reset this requires is not a workshop. It is a full recalibration of how leadership thinks about competitiveness. The question on the table is not "How do we use AI?" It is: "What kind of organisation do we need to become to remain competitive over the next decade?" That question belongs to the board. Every other conversation flows from how they answer it.

Part Two: Rewiring Decision-Making, Not Just Processes

Once the board owns the ambition, the next question becomes practical: What actually changes first?

The tempting answer is processes. Automate the routine work, accelerate the slow steps, reduce waste. This is the Industry 4.0 promise, and it has been underdelivering since 2016. Not because the technology was wrong, but because it was solving the wrong problem. Faster processes that feed the same decisions to the same people through the same hierarchies produce efficiency gains, not transformation. Gains your competitors can replicate within 18 months.

The real leverage is in decision-making. Specifically, in pushing decision-making authority down to the people closest to the work, and giving them the information and the mandate to act on what they find.

Across my years in technology I noticed the organisations that moved fast, that pivoted without losing momentum, were not the ones with the best systems. They were the ones that had put genuine decision-making power in people's hands. They cut the red tape. They trusted that the operator who has spent 15 years listening to a machine holds knowledge that no dashboard can fully capture and that this knowledge deserves authority, not just acknowledgement.

AI makes this more important, not less. It can surface information, model options, and reduce the cognitive load of routine decisions dramatically. What it cannot do is replace contextual human judgement: the synthesis that comes from years on the factory floor, from knowing that a particular vibration pattern always precedes a bearing failure even when the sensor data has not technically crossed the threshold. That knowledge, combined with AI capability, is what I think of as the subject matter expert (SME) on steroids.

The SME on steroids does not just understand their slice of the process. They own it from beginning to end, which is rarer than it sounds. One of the sharpest observations from industry roundtables I have participated in is how many people in large organisations only understand their segment of a workflow, position L to N, without visibility into what happened before their handoff or what their output enables downstream. AI forces end-to-end thinking because an agent needs someone to define the outcome it is optimising for. You cannot do that from a local perspective.

The professional development implications follow directly. Performance targets that measure activity need to give way to outcome ownership. A production supervisor's six-month plan should include owning two end-to-end processes, embedding AI meaningfully across at least 20% of those workflows, and mentoring two colleagues from other parts of the business in what they have learned. The mentorship requirement is how excellence spreads without a central authority enforcing it. The goal of every centre of excellence (COE) should be to make itself redundant. To distribute capability so thoroughly across the line that the idea of a centralised team guarding the knowledge becomes unnecessary. That is not failure. That is proof the CoE worked.

Part Three: The Pod Structure and the Death of Central IT

Individual empowerment will eventually stall if the organisational structure keeps pulling people back toward silos. The design has to match the ambition. That is where the pod concept becomes genuinely transformative, and where one specific role changes everything: the forward deployed AI engineer.

The forward deployed AI engineer (FD-AI engineer) is not a consultant who builds a solution and leaves. They are embedded in the business unit, sitting alongside the SME, learning the domain, co-building the tools and agents the team needs to improve their processes. The SME brings operational depth. The FD-AI engineer brings the technical capability to translate that depth into working systems. Over time the SME absorbs AI capability and the FD-AI engineer absorbs domain knowledge, and the boundary between business and technology dissolves in exactly the way it needs to.

This has an implication most traditional manufacturers are not ready for: It signals the end of the centralised IT function as we have known it.

The standard model for decades has been a global IT hub, somewhere separate from operations, staffed with technical talent who respond to requests from the business. That model had virtues around standardisation and cost control, but it created a structural distance between technology and work that made genuine transformation very difficult. When AI capability sits in a different building, accessed through a ticketing system, it cannot respond at the pace operational knowledge needs to evolve. A team building a predictive maintenance agent cannot wait three months for a requirements document to be approved.

The shift that needs to happen is a full dispersal of AI capability into every business unit's value chain. FD-AI engineers who used to sit in a central hub now sit in the production team, the procurement function, the quality team. They share the same KPIs. They attend the same operational reviews. They are accountable to the same outcomes. IT does not disappear. It transforms. It stops being a cost centre at the edge of the business and becomes part of how every function creates value.

This is uncomfortable for many IT leaders because it changes the model that made them powerful. But the leaders who see through that discomfort will find something better: When technology lives inside the business and the business succeeds because of it, the people who enabled that integration become some of the most valued in the organisation.

Part Four: Planning for the Agentic Workforce

The conversation about AI agents in manufacturing tends to collapse into fear or euphoria. Fear that agents will replace jobs and hollow out the workforce. Euphoria that agents are cheap, infinitely scalable, and will solve what decades of human effort could not. Both framings are wrong.

Agents are real and becoming more capable quickly. Early AI-first organisations are already running more than ten agents per human employee, handling well-defined, repetitive workflows with consistency and speed that human teams cannot match economically. For a traditional Swiss or European manufacturer, the relevant starting points are bounded: equipment monitoring, routine reporting, standard inventory flows, quality documentation. Places where the decision logic is clear, the data is available, and the cost of error is recoverable.

The most common mistake is framing this as cost reduction. An agent capable of handling a bookkeeping function to audit standard is not a $2,000 purchase. By the time you have trained it on your specific processes, validated outputs against edge cases, built the governance around it, and committed to ongoing retraining, you are looking at a six-figure investment per function. That is not an argument against agents. It is an argument for honesty about what they require. The organisations that build agents well frame them as capacity for the next growth initiative: scaling a new product line faster, entering a new market without proportionally increasing fixed cost. The agent becomes an enabler of the top line, not a reduction tool for the bottom line.

The CEO who leads this well does not start with a technology vendor. They start a conversation across operations, HR, and technology about where agents create the most value and what the honest implications are for the people currently doing that work. Some roles will change fundamentally. Some will disappear. The version of leadership that says, "Here is what the future looks like, here is the capability we need to build together, and here is how we will invest in every person to get there," is the version that earns the trust needed to actually make the change. In a workforce navigating this degree of uncertainty, that trust is not a soft value. It is the only thing that makes the change survivable.

The Transformation, Assembled

It starts with the board owning the ambition, not delegating it, not outsourcing it to technology, but genuinely asking what kind of organisation they want to be in a decade and accepting that the answer requires structural change, not just digital investment.

From that reset comes the rewiring of decision-making. Power moves down. The SME on steroids owns outcomes rather than tasks, measured on business results rather than activity.

From that empowerment comes the pod structure: FD-AI engineers embedded in business units so that capability grows where the work is done, and the old model of a central IT function serving the business from a distance begins to dissolve.

From that structural shift comes an honest agentic roadmap: agents planned as a growth resource, not a cost play, with the workforce implications faced directly rather than deferred.

And running through all of it is the variable that Japan demonstrated 40 years ago and that too many European manufacturers have still not absorbed. The competitive edge was never the machinery. It was the collective belief that the person closest to the problem is best placed to solve it, and that their knowledge deserves investment and authority.

That belief is still the variable. AI can amplify it enormously. But it has to be there first. You cannot amplify what does not exist.

The organisations that will still be competing in 15 years are not the ones that ran the most pilots or deployed the most agents. They are the ones that understood, early enough to act on it, that transformation is a human challenge dressed in a technology costume. The board that internalises that, and builds an organisation around it, is the board that wins. Everything else is a very expensive experiment.

About the Author

Alex Dean is a technology leader, entrepreneur, and AI transformation practitioner with 30 years of experience across solutions architecture, CTO roles, and go-to-market strategy. He leads the manufacturing practice at Unit8, working with industrial leaders across Europe navigating their next wave of AI transformation. Combining deep technical fluency with an MBA and entrepreneurial background, Alex co-designs programmes that are credible in the engine room and coherent in the boardroom. He is a regular speaker on AI strategy, data architecture, and organisational transformation.

Email: alxdean@gmail.com

LinkedIn: https://www.linkedin.com/in/alxdean

CHAPTER 7

THE STICKY NOTE THAT CHANGED EVERYTHING

By Eric Dressler
AI Leader and Founder, Agentsium
Edmonton, Canada

Out of difficulties grow miracles.
—Jean de La Bruyère

I grew up in the Canadian Prairies, where the sky felt endless and the winters taught you resilience whether you asked for it or not. My father served in the Royal Canadian Mounted Police, steady, principled, and committed to doing the right thing even when no one was watching. My loving mother worked at the telephone company, and for much of my teens and into my twenties, she lived with schizoaffective disorder.

As a kid, I didn't understand the diagnosis. I only understood the confusion. Some days she was herself: warm, funny, present. Other days she was lost in thoughts that weren't reality, and I felt helpless watching her struggle with things I couldn't see or fix. Growing up in an environment where the ground could shift without warning taught me something early: Clarity is precious. Information matters. And not having answers can be its own kind of pain.

Those years shaped me more than I realized. They taught me that nothing is more important than your health and your family's health. And they taught me something else: People often suffer not because the problem is unsolvable, but because they don't have access to the right information at the right moment. A tool like AI, even in its simplest form, would have helped us understand more, navigate more, and feel less alone.

The Hardest Battle of My Life

Years later, when my first son, Kaiden, was diagnosed with brain cancer at 22 months old, that lesson returned with a force I wasn't prepared for. There is no manual for that kind of moment. No leadership experience prepares you for watching your child fight for his life.

People told me they had never seen a father try so hard. A doctor in the United States emailed me once to say exactly that. My own father said I was a good dad during one of the long hospital nights. It meant a lot. But the truth is simple: Nothing I have faced compares to what my son went through, and that experience taught me that I will do anything for my family.

In addition to the countless hospital treatments, we became unexpected experts in medical cannabis, hyperthermia therapy, and other nonconventional techniques. We spent hours calculating THC and CBD ratios manually for cannabis oil, crosschecking studies, digging through archived research, trying to make sense of a world that felt impossible to understand.

I remember realizing that part of the doctor's treatment plan rested on a small Australian study from a few years earlier with fewer than ten children and a reported success rate of about 60%. As a parent fighting for your child's life, that isn't the kind of statistical confidence you cling to. It was a reminder that we were operating on hope and faith, not certainty. And it left me thinking: There had to be a better way.

And here is the part that still stays with me: If AI had existed the way it does now, it wouldn't have changed the outcome, but it would have changed the experience. We could have calculated dosages in minutes instead of hours. We could have analyzed research instantly instead of digging through outdated archives. We could have asked questions without waiting for appointments. We could have felt less alone in the uncertainty.

That realization is one of the reasons I believe so deeply in democratizing AI. Not for business. Not for hype. But because I have lived what it feels like to need answers and not have them.

The Moment AI Became Personal

My first real "AI changes everything" moment came in 2023, shortly after ChatGPT launched. It didn't happen at work or through some dramatic productivity breakthrough. It happened at home.

What most people don't realize is that GPT had been evolving quietly for years. The technology itself wasn't new. What changed the world was the moment someone put a chat interface on top of it. The moment people could simply interact with it, everything shifted. Intelligence became accessible. Capability became conversational. And suddenly, anyone could use it.

Around that time, my wife, Ashley, had been struggling with low iron levels, and it kept showing up in her bloodwork. What caught my attention wasn't just the iron, it was the pattern. Over time, her lipid levels were also dropping, almost in lockstep with the iron. I couldn't explain it, but something about the trend felt connected.

I asked AI to help me explore correlations between iron, lipid metabolism, and fatigue patterns. It didn't diagnose anything. It didn't pretend to be a doctor. What it did was surface a relationship worth paying attention to: Iron plays a role in how the body absorbs and processes certain nutrients, and disruptions in iron can sometimes show up alongside changes in lipid levels. It wasn't an answer. It was a direction, and it gave us one less thing to worry about, since the pattern likely stemmed from the low iron levels. This was something her doctor hadn't mentioned. We brought it to her appointment. It mattered.

That moment changed everything for me. Because if AI could help me see something meaningful in my wife's bloodwork, something I never would have spotted on my own, what else could it help with?

It turns out: almost everything.

It started small. Helping our daughter with math homework. Meal plans and converting measurements while cooking. Analyzing documents in seconds. Helping our dog when she wasn't feeling well. Drafting legal documents that used to cost thousands. Planning meals. Understanding symptoms. Turning ideas into prototypes in hours instead of months.

And then it snowballed.

I started using AI dozens of times a day. If I didn't know something, I used to reach for a dictionary or Google it. Now I reach for AI. I caught myself many times not using AI at work and finally put a sticky note on my monitor that simply said: Use AI.

It became the turning point. The habit builder. The reminder that if something was important enough to think about, it was important enough to ask AI about. That sticky note changed my life more than any productivity system I had ever tried.

The Final Mile Problem

For most of my career, I lived in the world of enterprise data. Twenty years of dashboards, data warehouses, predictive models, data analysis, governance frameworks, and executive presentations. I led large teams at very large companies. And through all of it, I watched the same pattern repeat itself: AI was always kept in the hands of the technical elite.

If you weren't a data scientist or an engineer, you were told, sometimes directly and often subtly, that AI wasn't meant for you. I always hated that. Because I knew the truth: Some of the smartest, most creative, most insightful people in any organization were the ones furthest from the technical tools. They understood the problems. They understood the customers. They understood the opportunities. Yet they were locked out of the systems that could bring their ideas to life.

Even when they did get access, the final mile always got in the way. Brilliant insights died in conference rooms because turning an idea into something tangible required a small army: a data team, a design team, a development team, a budget, a timeline, and a dozen approvals.

The best ideas in the world don't matter if they can't be expressed clearly.

The "final mile" problem always reminded me of the Las Vegas Rock and Roll Nighttime Marathon I once ran as a fundraiser for mental illness in my mother's honor. I raised almost 20 thousand dollars, but I trained only up to 60% of the distance. For longdistance training, you're supposed to train to 80% of the final race distance, but a foot injury stopped me short. I knew I was in for it. And I was.

Thousands of us were crammed together near the Mandalay Bay, vibrating with prerace optimism. My friends and family were out there yelling like I was about to win the Olympics. Macklemore was on stage hyping everyone up, and for a brief, delusional moment I thought, "Yeah, I'm built for this."

The first 15 miles of the 26.2 were glorious, the kind of miles where you start imagining yourself as a runner. Then I hit the Strip on the way back, and suddenly the universe reminded me who I really was. The bus, the one that quietly scoops up anyone too slow so they can reopen the Strip, started creeping up behind me like a horrormovie villain. I heard the driver yell, "Hop on." Absolutely not.

There was no way I was getting swallowed by the "loser cruiser" in front of thousands of tourists, so pure panic became my energy gel for the final two miles. I didn't come this far to come this far. I finished the marathon.

Afterward, I joked that every demographic known to humanity passed me. Grandparents. Teenagers. People in full mascot costumes. People vomiting. People who looked like they had been tricked into running by a friend who said, "It'll be fun." At one point I'm pretty sure a grandma wearing jeans and flipflops breezed past. Maybe I was hallucinating from the pain and fatigue. It was humbling, painful, and honestly hilarious, like the universe was giving me a highfive and a wedgie at the same time.

The first half was under two hours, but the second half was almost four. But I finished. Perseverance got me there. The final mile nearly broke me, but it also taught me something: You can do everything right, train for months, prepare your mind and body, and still find yourself struggling at the end because the last stretch is always the hardest. And finishing is always a motivator.

That is exactly what the final mile in data and AI felt like. Organizations could have brilliant ideas, talented people, and mountains of data, but when it came time to turn insight into something real, like a dashboard, a model, a prototype, or a decision, they hit the wall. They stalled. They lost momentum. Not because they lacked intelligence or effort, but because the final mile required tools they didn't have.

Then everything changed.

When ChatGPT launched, something clicked for me instantly. Not because it could write or summarize, but because it could translate. It could take an idea in my head and turn it into a prototype, a workflow, a design, a business plan, a legal document, a pitch deck, or a product. The final mile was now achievable.

I used to take a week to create a PowerPoint presentation. Now it takes 20 minutes. I used to pay lawyers thousands for simple agreements. Now I pay nothing. If I need a dinner recipe based on whatever ingredients are in the fridge, I ask AI. If I need a workout plan, I ask AI. If I need to understand a contract, I ask AI. If I need to brainstorm a business idea, I ask AI. If I need to teach my daughter something in a way that makes sense to her, I ask AI. If I wanted to start a global data and AI platform, I'd ask AI.

The Birth of Agentsium

In 2025, I was head of data and AI at a multibilliondollar international energy company. Then the company was acquired by an even larger energy company. Most people would have taken a break or found another executive role. But the timing was too perfect. Twenty years of experience. A lifetime of navigating uncertainty. And a new technology that finally closed the gap between idea and execution.

I wasn't pivoting. I was stepping into the work I was meant to do.

Agentsium didn't start as a company. It started as a question: What if anyone could build what they imagine? Not just engineers. Not just founders. Not just people with budgets and teams. Anyone.

What if a teacher could build a tool for her classroom? What if a nurse could build a workflow for her clinic? What if a parent could build a system to support their child? What if a creator could build the next big platform without writing a line of code? What if the next Netflix, Uber, Amazon, or Spotify wasn't built in Silicon Valley, but in someone's basement, kitchen, or coffee shop?

That's the world I wanted to create.

Once I started building, I couldn't stop. I built 20 products in six months. Not because I'm special. Not because I'm brilliant. But because AI finally let me close the gap between what I could see and what I could build. And I realized something important: If I could do this, anyone could, if they had the right platform to access.

That's what Agentsium became: a place where people can build AIbased apps for anything they can imagine, for their work or personal life. A platform where you can build ten times faster, ten times cheaper, and ten times better than traditional methods. A marketplace where you can monetize what you create.

We are even building an app called HAULO on our Agentsium platform, which connects truck drivers with hauls that need to be moved. Something that would have taken years to build traditionally can now be done in weeks using agentic AI, software that can take instructions, break them into steps, and execute those steps autonomously. It's like having a team of digital workers who never get tired, never lose focus, and never stop until the job is done. This isn't science fiction. It's happening now.

The Human Multiplier Effect

People think AI is about automation or efficiency or replacing jobs. They're missing the point. AI is a partner, a second brain, a sounding board, a pattern detector, a clarity machine.

In my house, AI is part of daily life. My wife, who doesn't consider herself technical at all, uses it to help our kids with school, convert measurements while cooking, understand her health, plan meals, organize schedules, and explore questions she never would have asked before. She once told me, "I don't know how I lived without this." And she meant it.

Our daughter uses AI to learn math in ways that make sense to her. Our dog has benefited from AIassisted research more than once. I use AI dozens of times a day, sometimes hundreds. Not because I'm obsessed, but because it's the fastest path to clarity.

AI doesn't replace people. It expands them. It unlocks parts of us we couldn't reach before. It bridges the gap between imagination and execution. It turns individuals into teams and ideas into reality. And it does something even more important: It gives people access. Access to knowledge, clarity, creativity, opportunity, and tools that were once reserved for the technology elite.

The Coming Disruption

The real risk isn't that AI will take your job. The real risk is that someone using AI will outperform you so dramatically that it won't be a competition. The gap is widening. But this isn't a warning. It's an invitation. Because the people who embrace AI early will experience the greatest leap in personal capability since the invention of literacy.

The Invitation

So here is my invitation to you. The next time you find yourself thinking too long without an answer, pause. Notice it. And then do what I taught myself to do: Ask AI.

Build the habit. Build the muscle memory. Build the life you want. Curiosity is the gateway. Questions are the fuel. Clarity is always one prompt away. You don't need permission. You don't need credentials. You don't need a technical background. You just need curiosity and the willingness to collaborate with a new kind of intelligence.

The world changed the moment someone put a chat interface on GPT. Your world can change the moment you decide to use it. And if you ever forget, put a sticky note on your monitor, fridge, or bathroom mirror. Just two words. The same two that changed everything for me.

Use AI.

About the Author

Eric Dressler is the founder of Agentsium and a longtime leader in data, analytics, and artificial intelligence. Before launching Agentsium, he served as head of data and AI at several large corporations. With more than 20 years of experience, Eric has dedicated his career to democratizing AI and empowering people to build what they imagine, regardless of technical background.

Eric's work is shaped by personal experiences navigating uncertainty, health challenges, and the search for clarity. Today, he builds AIpowered tools and platforms that help individuals and organizations close the gap between ideas and execution.

Email: eric@agentsium.ai

Website: www.agentsium.ai

LinkedIn: https://www.linkedin.com/in/eric-dressler-56a37416/

CHAPTER 8

THE EVOLUTION OF B2B SELLERS IN THE AGE OF AI

By Barry Flanagan
Founder, AI-IQ Systems
Granger, Indiana

There is a conversation happening in every sales leadership meeting right now, and it goes something like this: "Are we going to be replaced by AI, or are we going to use AI to replace each other?"

Both framings are wrong. The fact that we keep asking tells you something important about where B2B tech sales is stuck.

I have spent 25 years in enterprise technology sales. I have carried quotas, run territories, built teams, and watched the profession change in ways I did not always anticipate. The change happening right now is different from everything that came before it. Not because AI is more powerful than previous technology shifts, though it is. It's because it does not just change what sellers use. It changes what sellers are. The sellers who understand that distinction are pulling away from the pack. The ones who don't are getting quieter every quarter.

The Shadow Job Nobody Talks About

Here is a number that should trouble every revenue leader: B2B sales reps spend somewhere between 70 and 80% of their time on work that does not directly generate revenue: admin, CRM updates, research, drafting outreach, preparing for calls, and following up on emails that never get answered.[1,2] The actual selling—the conversations, the discovery, the negotiation, the relationship work—takes up less than a quarter of the average rep's week. We have built an entire profession around a shadow job, and then we measure people on the 20%.

I watch sellers do this all the time. They come in early, they work late, they are genuinely busy from the moment they sit down. But busy with what? Open the browser. Research the account. Draft the email from scratch. Log the activity in the CRM. Check the news on the company. Start again on the next account. Forty accounts in, it is noon. Twelve emails sent.

The rep sitting next to them, using AI to handle the research, the first draft, the CRM entry, has sent 50 emails in the same window. Not templates. Highly researched, targeted outreach built on real account intelligence. That rep has spent the rest of the morning actually talking to buyers.

This is not a productivity edge. It is a quota gap that compounds every week. Eleven hours a week is what AI gives back to sellers who use it properly (Unframe.ai, 2026). Eleven hours. That is more than a full working day, every week, returned to actual selling. Over a year, that math compounds fast: 44 extra selling hours a month, more than 500 hours a year. Over 14 additional workweeks. Three-plus extra months of selling time, no new headcount, no miracle quarter.

The sellers getting those hours back are not working smarter in the abstract. They are in more conversations, touching more accounts, following up more consistently, and building more pipeline. All while the rep next to them is still drafting from scratch.

The False Binary That Is Costing Deals

The conversation about AI in sales has been dominated by a framing that does not hold up in practice. On one side, you have the relationship sellers who believe the human element is irreplaceable and AI is a threat to the soul of the profession. On the other, you have the automation advocates who think enough technology and enough process can remove the messy, unpredictable human element from the deal cycle entirely. Both are wrong, and both framings are costing revenue.

The relationship-first argument rests on something true. Buyers value human connection. Gartner predicts that by 2030, 75% of B2B buyers will prefer sales experiences that prioritize human interaction over AI.[3] Buying committees experience significant internal conflict during enterprise decisions, and the seller who can navigate that conflict, who can read the room, identify the real objector, give the champion language that survives the internal meeting, is doing something no software can fully replicate.

But here is what the relationship-first camp misses: You cannot build relationships at the scale modern enterprise sales requires if you are spending four hours a day on admin. The relationship is not in conflict with the AI. The AI is what gives you time to have the relationship.

The automation argument makes the opposite error. It treats the deal like a logistics problem. Enough data, enough sequencing, enough triggered emails, and the revenue follows. The teams that leaned hardest into pure automation found that it works until it stops working. Buyers have grown highly sensitive to the difference between outreach built on real understanding and outreach built on a sequence. When 73% of B2B buyers say they avoid vendors who send irrelevant outreach, that is not a deliverability problem.[4,5] That is a relevance problem. Relevance requires judgment, which requires a human.

The teams winning right now are neither of these things. They are hybrids.

The GTM Engineer: A Profile Worth Paying Attention To

A new profile is emerging in high-performing revenue organizations. The term circulating for it is "GTM engineer." I have heard it from revenue leaders, from sales consultants, from people building go-to-market teams at high-growth companies and on LinkedIn. But the label stuck because it names something real that was happening before anyone had a clean way to describe it.

The idea is simple: The best sellers right now are operating less like traditional account executives and more like engineers who apply tools deliberately so their judgment can work at a higher level. An engineer does not choose between the tool and the judgment. They use tools precisely so their judgment can operate at a higher level. A structural engineer does not do calculations by hand because it builds character. They use software for the calculation so they can spend their time on the decisions that software cannot make.

The GTM engineer uses that same logic in sales. AI handles the work it is genuinely better at: research, synthesis, first drafts, CRM hygiene, account monitoring, follow-up sequencing. Then the GTM engineer brings their own judgment, their experience, their read of the buyer's political situation, their ability to have a hard conversation at the right moment, to the work that still requires a person.

This is not a new personality type. I have known sellers who always worked this way, reaching for whatever tools gave them more time for the conversations that mattered. What is new is that the tools are now capable enough that any seller who learns to use them properly can get the same result.

The divide is not AI versus human. It is engineered sellers versus overwhelmed ones. Teams using AI-driven workflows are seeing quota attainment rates dramatically higher than teams that have not. That is not a marginal improvement. That is a structural gap that, once established, tends to widen. The seller who is three months ahead on pipeline coverage has options the struggling seller does not. They can be selective. They can walk away from bad-fit deals. They can invest more in the opportunities that deserve it.

What AI Should Handle and What It Should Not

Part of the confusion comes from treating AI as a single thing with a single use case. It is not. The question is not whether to use AI. The question is what to use it for.

AI does some things better than any seller, regardless of experience level. It can hold more information about more accounts simultaneously than any human working memory. It can draft faster, research faster, cross-reference data across sources faster. It does not get tired on the forty-third account of the day. It does not write a worse email at four in the afternoon than it did at nine in the morning.

Use AI for that work. Research and account intelligence. First drafts of outreach, proposals, follow-up emails. CRM updates and activity logging. Competitive analysis. Meeting preparation documents. Synthesizing call transcripts into next steps.

AI also does some things badly, and this is the part that gets underweighted. AI does not read a room. It does not notice that the CFO leaned back when the pricing came up. It does not know how to handle the moment when a champion goes quiet because they are managing internal pressure you cannot see yet. It does not know when to push and when to give space. It does not build the kind of trust that survives a hard implementation quarter.

The sellers getting this right are not using AI to have more conversations. They are using AI to free up capacity so they can have better conversations. Those are different things, and the distinction matters.

I have talked to dozens of sellers who describe the same pattern. Once they offloaded the shadow job to AI, they were talking to more buyers, but the quality of those conversations changed too. They came in more prepared. More context. Less mental energy burned on admin before the call.

The Relevance Problem

There is a second dimension to this shift that does not get enough attention. It matters as much as the time argument. Enterprise buyers have changed. The information asymmetry that used to favor the seller, where the rep knew things about the market and the product that the buyer could not easily access, has largely disappeared. Buyers arrive at conversations having already done significant research. They have read the reviews, talked to peers, watched the demos, and formed opinions before the first call.

What they have not done is integrate all of that information into a picture of their specific situation. That is where a seller still adds something irreplaceable. Not by presenting information the buyer could have found themselves. But by helping the buyer understand what the information means for their particular problem, their particular organization, their particular stakeholders.

AI makes that possible at scale in a way it never was before. A seller with good AI tools can walk into a conversation having synthesized everything publicly known about a company's situation, their competitors, their recent announcements, their likely internal pressures. Not in a way that feels like surveillance. In a way that makes the conversation feel genuinely tailored.

Seventy-three percent of B2B buyers say they avoid vendors who send irrelevant outreach.[6] The sellers who close that relevance gap are not the ones who have the best product. They are the ones who show up to every interaction having done the work to understand the buyer's world. AI makes that work faster and more thorough than it has ever been.

What This Means for How We Build Sales Teams

If the GTM engineer profile is where the profession is heading, and the data says it is, the implications for how revenue leaders build and develop teams are significant.

Hiring criteria need to shift. The question is no longer just whether a candidate can work a room or handle an objection. It is whether they can build a workflow. Whether they are curious enough to experiment with tools and honest enough to measure what works. Whether they think of themselves as someone who uses judgment and technology together, not someone who chooses between them.

Coaching needs to change too. Most sales training is still built around the conversation: how to open a call, how to handle specific objections, how to negotiate. That training is still valuable. But it addresses the 20% of the job that most reps do reasonably well. The bigger leverage point, right now, is in the 80% that gets less attention. How a rep prepares. How they prioritize their time. How they use the tools available to them to arrive at each conversation better equipped.

The quota gap I described earlier, between the rep who has reclaimed 11 hours a week and the rep who has not, compounds over time. By the second year, the difference in cumulative selling time is measured in months. That is not a training problem that gets solved with a better objection-handling module. It is a workflow problem that requires a different kind of development conversation.

The Longer Game

Most discussions about AI and sales focus on efficiency metrics and quota attainment numbers. Something important gets lost in that framing. Those matter. But they are not the whole picture.

The sellers who build durable careers in B2B tech sales over the next decade are the ones who use the time AI gives back to get better at the things AI cannot do. The reading of a room. The building of trust over multiple interactions. The judgment about when a deal is real and when it is not. The ability to have a direct, honest conversation with a buyer who is making a decision that carries real risk for them.

Those skills have always separated the top performers from the rest. What is changing is the baseline requirement. A seller who cannot manage the administrative side of the job will not have enough time to develop those deeper skills at scale. The floor is rising.

The tools available to sellers right now are genuinely useful in ways that earlier generations of sales technology were not. The CRM was supposed to help sellers. Mostly it became another thing to feed. AI, used correctly, gives time back. That is a different relationship with technology than most sellers have experienced.

The sellers who treat AI as a way to protect and extend the human work rather than replace it are not just going to hit more quarters. They are going to build the kind of careers that compound. More relationships, more deals, more seniority, more credibility. All of it built on the same foundation: showing up to every conversation prepared, present, and worth the buyer's time.

That has always been the job. The tools just got considerably more capable.

References

1. Stephanie Condon, "Sales Professionals Spend 70% of Their Time on Non-Selling Tasks—AI Can Help," *ZDNet*, 2024, https://www.zdnet.com/article/sales-professionals-spend-70-of-their-time-on-non-selling-tasks-ai-can-help/.
2. Simplus, "70% of a Sales Rep's Time Isn't Spent Selling—Here's How to Change That," *Simplus Blog*, 2023, https://simplus.com/70-of-a-sales-reps-time-isnt-spent-selling-heres-how-to-change-that/.
3. Leonard Klie, "Buyers Are Pushing Back Against AI," *CRM Magazine* (Destination CRM), 2025, https://www.destinationcrm.com/Articles/CRM-Insights/Insight/Buyers-Are-Pushing-Back-Against-AI-171784.aspx.
4. Demand Gen Report, "3 out of 5 B2B Buyers Prefer a Rep-Free Buying Experience: Gartner," *Demand Gen Report*, 2024, https://www.demandgenreport.com/industry-news/news-brief/3-out-of-5-b2b-buyers-prefer-a-rep-free-buying-experience-gartner/49802.

5. Walnut, "B2B Buyers Prefer a Rep-Free Experience," *Walnut Blog*, 2024, https://www.walnut.io/blog/sales-tips/b2b-buyers-prefer-rep-free-experience/.
6. Demand Gen Report, "3 out of 5 B2B Buyers Prefer a Rep-Free Buying Experience."

About the Author

Barry Flanagan is the founder of AI-IQ Systems and has spent 25 years in B2B enterprise technology sales. He is the creator of Buyer Intel Accelerator, a B2B sales intelligence platform for tech sellers. He writes on deal execution, AI in sales, and buyer psychology at linkedin.com/in/barryflanagan and GetBuyerIntel.ai.

Website: www.GetBuyerIntel.ai

LinkedIn: https://www.linkedin.com/in/barryflanagan/

CHAPTER 9

FROM DATA TO DECISIONS: MAKING AI WORK FOR REAL BUSINESS VALUE

By Stephen Gatchell
Partner and Head of AI Strategy
Boston, Massachusetts

Without data, you're just another person with an opinion.
—Dr. W. Edwards Deming

In this chapter, we will look at why so many AI initiatives stall after early promises and what it takes to turn models and data into decisions that create measurable business value.

Artificial intelligence has moved beyond hype cycles and proof-of-concept experimentation. It now sits squarely in the boardroom, positioned as both an opportunity and an expectation. Yet for many organizations, the promise of AI transforming data into measurable business outcomes remains frustratingly out of reach.

But in today's environment, even that is no longer enough. Organizations are not struggling to access data; they're struggling to turn it into decisions that drive real, accountable value.

For most outcomes, the challenge is not a lack of data or algorithms. It is aligning people, incentives, trust, and execution so models actually change how work gets done. When that alignment is missing, teams build impressive prototypes while the business keeps making decisions with intuition, legacy reports, and competing priorities.

The Illusion of Progress: Why AI Efforts Stall

Many organizations begin their AI journey with enthusiasm. Data lakes are built. Tools are purchased. Data science teams are hired. Early prototypes show promise and leadership starts to believe the hardest part is over.

Then momentum slows. What looked like a straight line from data to value becomes a maze of approvals, integration work, conflicting priorities, and unclear ownership. Models sit unused, business teams lose confidence, and leaders question ROI not because the idea was wrong, but because the organization never built a reliable path from experimentation to execution.

At the core is a disconnect where AI is treated as a technical initiative, when it is a business transformation problem. The work is redesigning decisions, roles, workflows, and measurement so AI can be used, trusted, and scaled.

Or put more bluntly, echoing Thomas Edison: "Vision without execution is hallucination."

Execution is where AI becomes a repeatable capability instead of a collection of pilots. It requires shared ownership, a production-ready technical path, and governance that moves quickly without taking unnecessary risks.

The Personas in Conflict

AI delivers value only when multiple stakeholders move together from idea to day-to-day operations. Each group is rational from its own vantage point which is exactly why misalignment is so common. Figure 2 represents key stakeholder alignment.

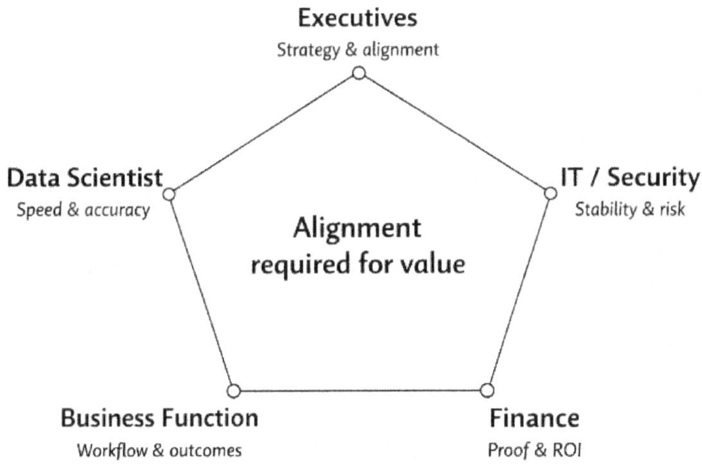

Figure 2

The Data Scientist: Speed, Exploration, and Model Quality

Data scientists iterate fast testing hypotheses, refining features, and improving accuracy. They need flexible data access, rapid prototyping infrastructure, and evaluation grounded in model performance.

That mindset is essential for discovery, but fast testing hypotheses do not run enterprises. Production demand's reliability, traceability, and change control, and "good enough" accuracy may still be unacceptable when decisions carry regulatory, financial, or reputational risk.

IT and Security: Stability, Governance, and Risk Management

IT and security are accountable for uptime, compliance, privacy, governance, and resilience. Their mandate is to ensure solutions scale safely and do not create fragile dependencies or unmanaged operational risk.

AI can feel disruptive because it introduces new tools, data flows, and vulnerabilities. Without clear ownership and guardrails, IT becomes a bottleneck, not from resistance, but from responsibility.

The Business Function: Urgency, Workflow Fit, and Outcomes

Business units adopt AI when it helps hit targets with less friction, better targeting, forecasting, or efficiency. Elegance does not matter; workflow fit and outcomes do.

That means value must be tangible and near-term, recommendations must integrate into existing workflows, and complexity must be hidden behind clear actions and trade-offs. Trust depends on transparency and how performance will be monitored.

Finance: Proof, Comparability, and Accountability

Finance ensures initiatives translate into measurable, attributable impact. They look for defensible assumptions and ROI that can be compared to other investments.

AI often struggles because attribution is messy as benefits can be indirect, shared, or delayed. Without an agreed baseline, measurement method, and ownership model, even strong implementations can be perceived as failures.

The Executive Team: Strategy, Alignment, and Operating Rhythm

Executives must align these perspectives into a coherent strategy and operating rhythm, deciding what to invest in and ensuring the organization can execute collectively.

In practice, which means setting ownership (who is accountable for outcomes), prioritization (which use cases matter most), and consistency (how models are built, deployed, monitored, and retired). Without coherence, AI is everyone's priority in decks and no one's priority in operating meetings.

Where Value Gets Lost

Value is often lost in the handoffs: (1) experimentation to production, when prototypes can't be integrated, supported, or owned; (2) access to governance, when speed and controls aren't balanced; (3) insight to action, when outputs don't become part of a workflow; and (4) impact to measurement, when nobody agreed how to quantify and attribute value. Figure 3 reflects the pipeline and the potential cracks that leads to value leakage.

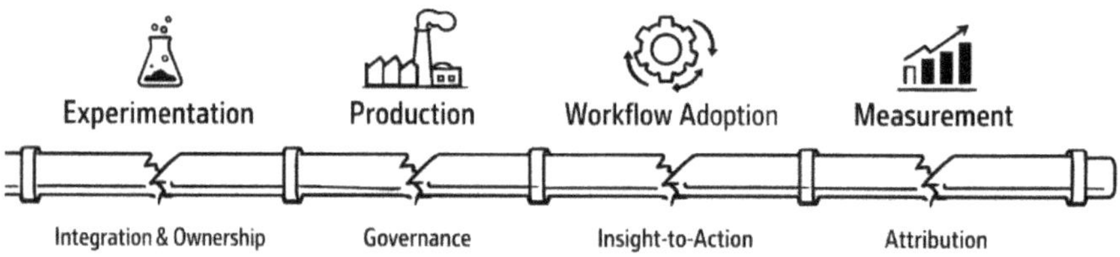

Bridging these gaps requires disciplined lifecycle management and explicit ownership. Data definitions must be managed like products. Models must be monitored for drift and changing conditions, with playbooks for retraining or rollback. And operating models must clarify who maintains pipelines, who signs off on risk, who drives adoption, and who owns business outcomes.

Case Study 1: The "Perfect Model" That Delivered No Value

A large enterprise built a highly accurate churn model that identified at-risk customers with impressive precision. Six months later it still was not in production. Integration was never planned as a first-class deliverable, and no one defined how a risk score would trigger an operational action.

IT raised valid questions about deployment and support, marketing lacked clarity on what to do differently, and there was no accountable owner to drive adoption. The technically successful project produced zero business value.

Lesson: Accuracy does not equal impact. A model creates value only when it reliably changes decisions and behavior in the business.

Case Study 2: Aligning Around a Financial Target

Another organization started with a financial question: "Where can we create $10 million in measurable value using data and AI?" That reframed the work from possibility to accountability.

Cross-functional teams (data, IT, business, and finance) identified use cases and evaluated each on feasibility, impact, and measurability. Finance defined baselines and clarified how value would show up in reporting.

One pricing initiative clarified the decision to improve, defined success metrics tied to revenue and margin, established a baseline, and then shipped incremental improvements that could be tested and expanded. Because measurement was designed in parallel with delivery, leaders could credibly claim impact as soon as results appeared.

Within months, the organization could attribute millions in incremental revenue to the initiative because the measurement approach was agreed upfront and because the solution fits how the business actually operated.

Lesson: Start with value, not technology. When teams align on the measurable outcome, they become motivated to solve the operational details that make AI real.

Case Study 3: The Cultural Barrier

A third company had modern platforms, skilled teams, and executive support, yet still struggled to scale adoption because of culture and trust, not technology.

Business teams hesitated to trust recommendations; data literacy was uneven; change management was minimal; and earlier analytics efforts had created skepticism. Leaders also worried about accountability: If a model drove an action that went poorly, who owned the outcome?

They responded with role-based training, clearer explanations of how models should be used, and they embedded data scientists in business teams. Lightweight responsible AI practices (purpose/limits, bias checks, and escalation paths) improved trust and adoption followed.

Lesson: Adoption is about people, not just technology. Trust is built through usability, clarity, and visible accountability.

Making AI Work: A Practical Framework for Real Value

Organizations that turn AI into measurable value share a few consistent principles. The difference is discipline: treating AI as a lifecycle capability that is funded, governed, adopted, and measured like any other strategic investment.

Anchor the work in decisions and outcomes. Start with the decision to improve, the metric it influences, and the financial value to capture. If you cannot name who makes the decision, how often, and under what constraints as well as defining a baseline, then you will struggle to prove impact even if the model "works."

A useful test is whether you can state the initiative in one sentence with a verb and an owner: "The Chief Customer Office wil be accountable to reduce churn by prioritizing offers for the top 5% of at-risk accounts weekly," or "the Chief Revenue Officer will iimprove margin by recommending price bands at the point of quote."

Bring finance in early and keep them close. Finance should be a partner from the beginning defining how value will be measured, validating assumptions, and agreeing on attribution that will stand up to scrutiny. Early involvement drives better design choices: instrumentation, baselines, and credible comparisons. It also forces governance clarity: realized versus theoretical value, who owns P&L

impact, and what evidence is required. That discipline increases leadership confidence to scale what works.

Design for adoption, not just accuracy. A model that is not used has no value. Design for adoption from day one: embed recommendations in existing workflows, present them the way decisions are made, and make the next action obvious.

Treat adoption as a success metric alongside model quality, track usage, override rates, cycle time, and downstream outcomes. Build simple feedback loops so users can flag bad recommendations and improve the system with real-world input.

Align incentives and clarify ownership. A common barrier is misaligned incentives paired with unclear ownership. When leaders optimize different objectives including growth, cost, and risk, collaboration slows. And when "everyone" owns AI, no one owns it.

Successful organizations make shared outcomes explicit in goals and operating reviews. They treat high-value use cases like products: A named business owner is accountable for outcomes; product/ops roles drive adoption; engineering owns reliability; and risk teams provide clear guardrails.

Build cross-functional teams that own outcomes. AI can't live in a silo. Effective teams combine data science, engineering, business stakeholders, and finance from the start, engaging IT/security early to shape a deployable solution and staying accountable through deployment, adoption, and measurement.

Invest in data literacy, change management, and responsible AI. AI literacy is no longer optional. Leaders need a strategic grasp of what AI can and cannot do, users need practical guidance on using recommendations, and technical teams need decision context. Without shared language, organizations swing between overconfidence and skepticism.

Change management is equally critical. People must understand what is changing, why, and how accountability works when human judgment and machine recommendations intersect. Responsible AI becomes practical here to document purpose and limits, protect sensitive data, meet regulatory requirements, and define review processes for bias and explainability.

Start focused, prove value, then scale through a repeatable factory. Large, multi-year programs often fail under their own weight. Start with a focused use case that matters, prove value quickly, and scale the organization's ability to deliver repeatedly reusing patterns for pipelines, testing, deployment, monitoring, and measurement.

That repeatability is the "AI factory": shared standards, tools, and governance that reduce the cost and time of each use case. It includes core MLOps (versioning, automated tests, deployment, monitoring, retraining triggers, incident response) and clearer lifecycle costs. It also sharpens build-versus-buy choices: Vendors can accelerate delivery but add dependency; internal builds offer control but require sustained investment. Closed versus open-sourced models should be evaluated to determine the best model for the defined outcome of the use case with clear understanding and alignment for the benefits of each.

The Measurement Challenge: Attributing AI Value

Attribution is one of AI's hardest problems: How do you prove a model drove an outcome versus an external factor? If you cannot measure impact credibly, you can't prioritize, fund, or scale AI with confidence.

The strongest approaches use A/B tests, controlled experiments, and baseline comparisons. When that's not feasible, phased rollouts, matched cohorts, or time-series analyses can still build credible evidence if limitations are explicit. The key is designing measurements from the start, including instrumentation to track recommendations served, accepted, or overridden, and downstream outcomes.

Most important is agreement upfront: data sources, baselines, time horizon, and how confounding factors will be managed. Without it, real value gets disputed and teams argue instead of improving.

The Role of Leadership: Driving Alignment

AI success ultimately comes down to leadership: choosing outcomes that matter, breaking down silos, and insisting on accountability. Leaders also decide whether governance accelerates progress or slows it down.

High-performing organizations create a clear cadence: a regular forum to prioritize use cases, review risk, track adoption, and measure value. Decision rights are explicit, which prevents the most common failure mode which is indecision. Consider making explicit the following decision areas:

- Define principles.
- Set clear expectations for value.
- Break down silos between teams.
- Be explicit about roles and responsibilities.
- Align incentives and priorities.
- Foster a culture of experimentation and accountability.

This is not easy. It requires balancing speed with control, innovation with risk, and autonomy with alignment.

A Final Reflection: From Capability to Impact

Many organizations have already built the capabilities required for AI including data platforms, tools, and talent. Yet capability alone does not create value.

The journey from data to decisions is not linear. It is iterative and collaborative, and it demands alignment across teams. Organizations that build that alignment unlock something powerful: AI not just as a capability, but as a driver of measurable business impact that aligned with the corporate strategy.

Closing Thought

The question is no longer whether AI can create value. The question is whether organizations can align themselves to realize it. Those that do will not just adopt AI but will transform with it. Because in the end, having data is not the advantage; acting on it is.

Key Takeaways

- Align on principles across key functions.
- Treat AI as a business transformation effort, not a technical side project.
- Define the decision to improve and the metric to move before building models.
- Design for adoption: Integrate output into workflows and make recommendations actionable.
- Bring finance in early to agree on assumptions, attribution, and measurement.
- Cross-functional teams and aligned incentives are the fastest path from prototypes to production impact.
- Be explicit with roles and responsibilities.
- Invest in literacy and change management to build trust and sustained usage.

About the Author

Stephen Gatchell brings a unique perspective on data and AI shaped by three roles: enterprise leader of large-scale enterprises, advisor from a vendor perspective, and now a consulting partner. Stephen has led global teams across data, analytics, governance, and AI; advised hundreds of organizations on modernizing data foundations; and now delivers practical, scalable, outcome-driven solutions. This blend of first-hand and consulting experience enables Stephen to move clients beyond buzzwords to significant business value.

Having navigated the operational realities of enterprise data and AI, Stephen has been accountable for strategy, platforms, data and AI products, and transformation initiatives. Stephen has advised CxOs on data, AI, privacy, and security while helping business leaders align data and AI with measurable outcomes.

At Ortecha, Stephen drives the AI strategy and helps organizations simplify complexity, unify data efforts, and build sustainable, automated, and reliable AI-driven ecosystems that align to corporate strategies.

Email: Stephen.gatchell@gmail.com

LinkedIn: https://www.linkedin.com/in/stephengatchell/

CHAPTER 10

EVERYONE HAD A PLAN

By T. Alexander Gorrie
Founder and CEO
London, England, United Kingdom

Everyone has a plan until they get punched in the face.
—Mike Tyson

For most of my career, the technology industry had a plan. It was not written down in one place, but everyone understood it. Companies bought software designed by someone else. Those systems were implemented across the organization, and teams adjusted their processes to fit the structure of the tools they had purchased. If something did not quite work, the expectation was simple. You adapted.

At the time, it never felt like a plan. It felt like reality. You chose the best software available, and in many cases my team or I was helping companies buy it. Organizations assembled stacks of tools designed to solve specific problems and then spent considerable effort trying to make those tools communicate with one another.

Sometimes the systems connected. Often they did not communicate particularly well. Over time, people accepted the friction that came with this approach. It was part of the job.

For most of the past two decades this was the world of software as a service (SaaS). SaaS promised efficiency at scale, and in many ways it delivered. Companies adopted platform after platform in the pursuit of productivity and coordination. Entire industries were built around helping organizations select, implement, and manage these systems. The model worked well enough that very few people stopped to question it.

Yet beneath the surface there was a quiet contradiction. The technology was powerful, but rigid. Organizations, by contrast, were dynamic and constantly evolving. Processes changed. Teams reorganized. Strategies shifted. The software, however, remained largely fixed in the structure designed by its creators. The gap between those two realities was where most of the real work happened.

Over time that gap became surprisingly large. The average company now operates hundreds of software applications across the organization. Large enterprises use an average of more than 300 apps. Yet the investment is often inefficient. Research shows that more than half of SaaS licenses go completely unused. Even when employees log in, engagement is shallow. Industry benchmarks show average core feature adoption around 24%, with a median closer to 16%. In practice this means that most organizations use only a small fraction of the capabilities they purchase. An easy way to think about it is this: How many apps on your phone do you actually use?

At the time, this arrangement felt normal. Businesses optimized within the model rather than questioning the model itself. Teams became skilled at adapting their processes to the structure of the software they had purchased. If a workflow felt awkward or inefficient, it was usually treated as an operational problem rather than a technological one.

The Industry Had a Plan. Then Something Changed

I do not remember expecting much the first time I experimented with artificial intelligence. It felt like just another tool. Another interface layered onto a stack that already contained more systems than anyone truly needed. Most of us had seen many waves of technology promising efficiency before. There was little reason to believe this one would be fundamentally different.

Within minutes, however, something felt unfamiliar. Instead of navigating a predefined system, I was describing what I wanted the system to do. Instead of adjusting my thinking to fit rigid workflows, I could articulate the outcome I was trying to achieve and watch the system begin to assemble it. The results were not perfect. They were sometimes rough and incomplete. But they were clear enough to reveal something important.

This was not simply better software. It was a different relationship to technology. It was the moment the plan stopped working.

For most of the SaaS era, software dictated structure. Companies that tried to bend a platform to their exact needs often struggled, while those that reorganized their processes around the software generally succeeded. Product teams designed workflows that worked well for a broad set of customers, and organizations inherited those assumptions whether they fit perfectly or not. Implementation became the art of aligning human behavior with the structure embedded in the technology.

The language of the industry reflected this reality. We spoke about best practices and digital transformation. We invested in change management and training programs designed to help teams adopt the tools they had purchased. Beneath all of this activity was a quieter truth. Organizations were adapting themselves to the limits of their software.

Artificial intelligence begins to reverse that dynamic. For the first time, useful software does not need to be fully predefined before it can deliver value. Systems can be shaped in real time through context, instruction, and iteration. The implications of this shift are easy to overlook at first, but they are significant. The concept of an average user begins to lose meaning. The idea that a single platform should serve every organization in the same way becomes less convincing. Software begins to feel less like a static product and more like a system that can evolve alongside the people using it.

I began to understand the full weight of that shift after leaving my previous role. For years I had worked inside the SaaS ecosystem, helping companies buy and implement software built by someone else. I understood how the system worked. I also understood its limitations. Eventually the timing felt right to try something different.

There were no existing workflows to improve. No internal systems to optimize. The platform that would become Dromeda.ai did not exist yet. The only thing that existed was the idea.

Historically, building a software platform from scratch required a familiar path. You assembled an engineering team. You defined a development roadmap. Months passed before the first version of the product was usable. This time the process looked very different.

Instead of writing thousands of lines of code manually, I began describing the system I wanted to exist. I outlined how the platform should behave, how information should move through it, and how users would interact with it. The AI generated code, suggested structures, and helped translate ideas into working components. I reviewed the output, refined the instructions, and adjusted the architecture. Then I iterated.

The development process felt less like traditional programming and more like guiding the creation of a system. Today people often refer to this approach as vibe coding. The phrase did not exist when I started building the platform. It emerged later as more developers began describing similar experiences working with AI-assisted tools. The label came afterward. The shift itself was already underway.

For someone who had spent years helping companies purchase software designed elsewhere, the experience was striking. The distance between an idea and a functioning system had suddenly become much smaller. Once you experience that collapse in distance, it becomes difficult to see technology the same way again.

Software stops feeling like something organizations purchase and implement. It begins to feel like something they can shape.

The economics of software begin to change as well. For many years the value of enterprise software was tied primarily to access. Companies paid annual subscriptions and per seat licenses in order to use the capabilities embedded in the platform. Pricing models reflected the scarcity of the technology itself.

Artificial intelligence challenges that assumption. Access to the technology matters less than the outcome it produces. Organizations begin to measure value in terms of time saved, costs reduced, and decisions improved. Software becomes less about the product that is purchased and more about the leverage it provides.

This shift also alters the role of the individual inside an organization. The people experimenting with AI are not simply working faster. They are approaching problems differently. Tasks that once required coordination between multiple teams can sometimes be automated or redesigned by a single person. Small systems can be built to replace complicated processes. Ideas move more quickly from concept to execution.

Perhaps most interesting is the fact that many of the individuals doing this work are not traditional engineers. For decades the ability to build software was gated by technical expertise. That gate is beginning to weaken. What matters increasingly is not the ability to write code from memory, but the ability to understand a problem clearly and guide a system toward a solution.

This shift inevitably creates tension. Technological change always redraws the boundaries between those who adapt quickly and those who continue operating within older assumptions. Some individuals experiment and build new capabilities. Others wait for established tools and processes to catch up.

History offers many examples of this pattern. The printing press did more than make books cheaper. It changed who could access information, who could distribute it, and who ultimately influenced society. Artificial intelligence appears to be creating a similar dynamic. The details are different. The pattern is familiar.

For years the industry had a plan for how software should work. Artificial intelligence simply exposed how fragile that plan had become.

For many years I believed that progress depended on finding better tools. Looking back, the deeper shift was not about tools at all. It was about how we thought about them. I thought I needed better tools. What I needed was to stop thinking like someone who did. The real question now is what people choose to build.

About the Author

T. Alexander Gorrie is the founder of Dromeda.ai, where he builds AI-native systems designed to improve the commercial viability of medical technologies. With a background in enterprise SaaS and healthcare, he has spent years helping organizations scale both software and medical diagnostics. His work now focuses on redefining how software is created and used in an AI-driven world, with a particular emphasis on advancing healthcare innovation through commercial viability.

Email: alexander.gorrie@gmail.com

LinkedIn: https://www.linkedin.com/in/alexandergorrie/

CHAPTER 11

THE AMPLIFIED HUMAN

By Christopher Hartline
AI & Cloud Architect
Lenexa, Kansas

Art is the signature of man.

—G.K. Chesterton, *The Everlasting Man*

OpenAI's "ChatGPT moment" was not expected, but it could have been. If you look at the progress that was being made, the general direction of travel, you could have anticipated the change.

The AI ground floor wasn't 2019.

Some might pin the date to 2022 with the release of generative AI products such as Stable Diffusion, Claude, or OpenAI's ChatGPT?

I believe it's now, or 2025 with the release of second-generation foundation models like Claude Sonnet 3.5, agents and the arrival of the agent skills and the model context protocol (MCP). I'm saying this to encourage you to get involved. Now.

Development over the last 18 months has been straightforward with few surprises. Models have improved. Context windows have grown. Foundation model providers extended their original language models to include multimodal capability and next-generation libraries. Text-to-image and video via diffusion systems have improved, but we've been refining existing technology. That means it's reasonable to draw out what could be, based on what's been.

Beginning Now

Language models emerged from research that began in the mid-2010s at Google. Technologies like transformers and vector embeddings proved foundational. They gave machines the ability to receive input in natural language, process it, and respond in kind.

The paper "Attention Is All You Need," published in 2017, introduced multi-head attention, the architecture beneath every modern large language model. That same year, AlexNet had already demonstrated what was possible with deep learning in image recognition. In 2012, it had decimated previous benchmarks in the ImageNet competition, dropping the error rate from 26% to 15% and signaling that neural networks had crossed a threshold. A year after the transformer paper, Google released BERT (bidirectional encoder representations from transformers) in 2018. Here's where things pick up speed.

Companies like Anthropic and OpenAI built on that foundation, refining early encoder-decoder systems into the products we recognize today: Claude and ChatGPT, both released in 2022.

But the models didn't emerge in isolation. They rode a wave of converging infrastructure.

Kaggle launched in 2010 as a data science competition platform and was acquired by Google in 2017. Over time it evolved to welcome newcomers through tutorials and shared datasets. Today it hosts AI and ML models alongside its original community of practitioners.

HuggingFace was founded in 2016 and pivoted to open-source machine learning in 2019 with the release of their Transformers library. Today they're best known for their model registry, their open-source libraries, and Spaces, managed environments where anyone can deploy AI models for little to no cost.

All of these brought interest, onboarding thousands of people into the AI and ML community. But model hosting and the models themselves had to mature over that same time frame.

First, cloud maturation. From roughly 2006 through 2018, the major cloud providers—AWS, Google Cloud, Azure—grew from experimental platforms into reliable infrastructure. The tools were still clunky in the early years. For most organizations there was little reason to get excited.

Second, deployment. As the cloud matured, communities and companies began moving workloads in. For me, that shift began in earnest around 2018. The SDKs improved and friction dropped. This is when acronyms like SaaS (software as a service), PaaS (platform as a service), IaaS (infrastructure as a service), and IaC (infrastructure as code) became common.

Third, AI maturation. By 2023, foundation models crossed a threshold. Claude 3.5 Sonnet is often cited as an inflection point. Its ability to maintain large context windows and session coherence made new kinds of applications possible. Agent frameworks like LangGraph appeared. Tool-calling became reliable. The building blocks for autonomous systems fell into place.

All along, a parallel shift was underway: a growing willingness to adopt open-source tools for commercial use. Linux and Apache had been staples for a decade. But Docker (for containerization) and Kubernetes (for orchestration), both released by Google to the community, changed how software got deployed. And while there are no open-source cloud platforms, there are thriving open-source communities for datasets and AI models: HuggingFace and Kaggle. If you're in the cloud, your provider likely offers them as managed services.

Here's Why These Matter

You can rent time on an Nvidia A100 GPU through Google Colab+ for less than $50 a month. The GPU itself costs over $12,000, if you can get one. That doesn't include the infrastructure and power to run it. With that access, you can fine-tune a foundation model in hours using techniques like parameter-efficient fine-tuning.

Let me make this concrete. Yesterday morning I downloaded a Llama 3.2 8B model and a conversational dataset from HuggingFace. Ten minutes. I spent another 30 minutes working with Claude Code to build a fine-tuning script. Then I trained the model on Colab+ for five hours.

That evening, I had merged the output files and was running the fine-tuned model locally on my laptop. The next day I added memory, a system prompt to shape its personality, and retrieval-augmented generation (RAG) so it could pull internal documents. Part-time effort, spread across two days. The result: a special-purpose, locally hosted AI agent built on an open-source foundation model. No internet required to operate once built and trained. Compute, data, community, tools all accessible, all affordable, all assembling into something that wasn't possible five years ago.

AI as a Collaborator

Please hold two seemingly contradictory thoughts in balance: *Human skills are becoming more valuable, and AI consumes any problem that can be digitized.*

Here's what I'm asking you to consider. AI lacks creativity, experience, and judgment, but it provides breadth, speed, and precision. Creativity isn't quick, and human experience is generally narrow. When operating together, human-AI collaboration enables rapid innovation because the individual closest to the problem can now play an active role in the solution. This is where human skill and experience become more valuable.

Andrej Karpathy has said that the hottest programming language is English. Coding via a prompt has been derided as vibe coding, and perhaps some of that scorn is earned. But it's only going to improve. Rapidly.

The natural progression has become a process where an individual uses their intuition and experience to create a general outline of the problem and an acceptable solution while the AI assistant retrieves information and generates a plan.

Today, the domain expert that's closest to the problem can address it without needing assistance from a team of technical and operations experts. The domain expert's AI assistants can fill those roles. All of them. *Whatever you're doing, AI is additive.* If you're a small-business owner, AI can provide legal recommendations. If you're a lawyer, AI can be an accountant. If you're an accountant, AI can be a business assistant.

This is my belief and personal experience. Just this morning I came across an article highlighting this exact occurrence—lawyers, doctors, and construction workers winning hackathons and delivering differentiated solutions by combining their professional knowledge and experience with an AI assistant.

Be the domain expert with an AI assistant. Maybe even one you create.

Amplifying the Domain Expert. Augment, Don't Replace.

Let's extend the idea above. Tool use, memory, valuable data, and reasoning make AI assistants more useful.

If artificial intelligence is software meant to mimic human decision-making, then an AI agent is an autonomous AI system given tools and agency, configurable abilities and a defined scope of operations.

It's important to recognize that while it's common for people to get concerned when told that AI has tools and is autonomous, this isn't new. What's occurring now is a refinement of what's been happening for years. Robo-dialers, spam-messengers, and IT infrastructure have been automated for close to a decade.

Software as a service, cloud provider managed services, and enterprise IT operations have used automation (Jenkins, Puppet, Ansible, ArgoCD, Kubernetes, Terraform, PowerShell scripts) as a centerpiece to the product, service, or operating capability.

In fact, one of the first agents I built was a system administrator. Linux documentation is all over the internet, so it's well represented in every foundation model's training base. Likewise for Cisco and Windows PowerShell. My agent did four things: received a query, reasoned over the query, called tools, and responded to my query with commands and scripts.

My goal was to create an AI agent that could assist system admins and site reliability engineers. This is beneficial for senior folks but very useful for new, junior personnel. Senior folks have a reminder; junior personnel have a tutor.

If we adapt the Google ML maturity model for our use with AI, this is a tier-one capability, something that's partially automated.

Let's raise the stakes by including infrastructure as code. A user might have an AI assistant that can use a Terraform tool to create, manage, alter, and deploy Terraform patterns. This can be done safely, or not. A safety guardrail should absolutely require human approval before any destructive operation. After all, the biggest threat is the insider threat, right? Anyone who's been in IT for more than six months knows is familiar with that concept.

Let's pivot to a more benign case: the research assistant. It's the AI engineer's "Hello World." It's such a popular use case that all agent frameworks have prebuilt libraries to enable it. A research assistant agent is usually a multimodal model that receives queries from a user, uses tools (web search, document retrieval) to retrieve information, processes the information, and provides a response. This is the same pattern used above to create a virtual system administrator.

Here's the path I'm trying to illuminate: We began with foundation models; later we gave them tools. We created agents. We matured agents. Now there's a movement to standardize the deep research agent pattern. *Capability is accumulating such that custom code you or I might need to create next week could become a common library in six months.*

As an adjacent example: Did you know that there are specialized medical and financial large language models? And since they're almost certainly based on contemporary open-source models, there's no reason you shouldn't be able to give them tools, maybe access to email or messaging and APIs for financial market data providers.

"Dark Factories"

There are "dark factories," software development organizations where developers don't code. One hundred percent of the software is developed by AI. Humans focus on identifying the problem, designing instructions, and running evaluation of the AI model's output.

Just as a physical factory can run with the lights off because robots are doing the assembly, software organizations are beginning to move toward a tier-two maturity model:

- People identify the business-need and define success. They articulate the problem and set the goal (e.g., preferred outcomes). The art of design thinking.
- AI generates the code, runs verification testing, and deploys to staging. The focus here is on "building it right." Building to spec.
- People perform validation testing. The focus here is "Did we build the right thing?" Is the product or service useful as built? Is value provided?

This isn't the replacement of the developer; it's empowering the developer with a new tool. The speed of innovation is no longer limited by how fast a human can type or debug code, but by how clearly they can think.

Organizational Intelligence and Institutional Memory

To harness the power of agentic AI, an organization only needs to do two things: identify the problem and understand their business.

AI won't save an organization that can't define its problems or document its processes. But for organizations that do have that clarity, something powerful can occur.

Another consideration is brain drain and the knowledge that walks out the door. In the legacy model, we attempted to solve this with documentation: standard operating procedures, hand-over folders, shared drives. Documentation is static and usually stale. It doesn't capture the *why* behind a decision or the nuanced judgment of a senior architect who's seen three similar projects fail.

We're moving toward something different:

From search to synthesis. Institutional memory can be a conversation based on recent, relevant information instead of a keyword search. This is enabled via RAG (retrieval-augmented generation). With RAG, an organization can ground a foundation model in its own private data, wherever it resides. And since agents can call tools and provide structured output, it's almost as easy to update documents and knowledge bases as it is to read them.

Capturing what never gets written. The most valuable knowledge is often ephemeral. As AI becomes integrated into development and operational cycles, it accumulates the organization's collective judgment.

From knowledge to action. This leads to the integration challenge. For AI to truly serve as institutional memory, it can't be an isolated chatbot. It must be woven into the fabric of work.

Model context protocol (MCP) is an emerging standard that provides AI agents with a universal connection pattern to data, APIs, and internal tools. It's a USB for your AI agent. An organization can build MCP servers that expose data, resources, and functions to its AI systems, creating shared infrastructure for enterprise-wide agents.

Let's revisit our Terraform example. Now our agent knows *why* a certain Terraform module was built, why it was designed the way it was. The agent that can update that module based on that history—while citing the original human logic—is revolutionary.

Patterns

We went from mainframes to personal computers, laptops to the cloud, and then to handhelds - tablets and phones. On the software side we went from locally developed and run applications, to commercial applications, then apps.

I fully expect AI to take a similar path. It'll migrate out of the data center. Foundation models will stay. They'll be used to create personalized AI: your AI agent. It'll be fit-to-purpose and local, with "apps" becoming agentic tools. And you'll be working with them in natural language. First via text, later via spoken language. No doubt this will be big business. Maybe yours? Or start learning now and build your own.

Final Thought

Classical computers process information in bits—ones and zeros, on or off. Quantum computers operate on fundamentally different principles. They exploit properties of quantum physics that have no analogy in everyday experience: superposition, entanglement, the ability to explore many possibilities simultaneously rather than sequentially. This isn't incremental improvement. It's a different kind of machine.

The AI systems transforming work today run on classical hardware. There are technical and material reasons to consider alternatives, hybrid-classical quantum systems.

Researchers are exploring how quantum properties could accelerate the core operations of neural networks: optimization, search, pattern recognition across high-dimensional spaces. The intersection of quantum computing and AI isn't speculated, it's active research. Actually, there is a lot of material on this topic for optimization and inherently quantum processes with interest in quantum neural networks (QNN) increasing.

No one can tell you when quantum-enabled AI will arrive at scale. The timeline remains uncertain. But the pattern should look familiar by now: technologies converging, infrastructure maturing, access democratizing. We've seen this before. We're living the result.

The organizations and individuals building AI fluency today are developing the "cognitive muscle" and collaborative instincts that will matter when the next convergence arrives. The skills transfer. The mindset transfers.

We all have a window into the wonderful future. Quietly. Then suddenly. It's going to happen again. You've got this.

About the Author

Christopher Hartline is a Catholic husband to Jennie and father to Hope, Isabelle, Audrey, and Anna. He's a doctoral student and principal architect with expertise in cloud, AI, and blockchain.

Email: chris.hartline.99@gmail.com

LinkedIn: https://www.linkedin.com/in/chrishartline/

CHAPTER 12

AI WON'T REPLACE YOU, BUT PEOPLE WHO USE IT WILL

By Stephan Heilmann
AI Leader and Advocate
Zug, Switzerland

The illiterate of the 21st century will not be those who cannot read and write, but those who cannot learn, unlearn, and relearn.

—Alvin Toffler

Whenever a technology arrives that genuinely rewrites the rules, people reach for the same tired questions. They asked them about the steam engine, electricity, the internet, and now they are asking them about artificial intelligence: Will it take my job? Is it dangerous? Which product should I buy? These are reasonable questions, but they are the wrong starting point. They treat AI as a gadget to be evaluated rather than a force to be harnessed. And that misunderstanding—that AI is primarily a technology story—is exactly what separates the people who will flourish in the next decade from those who will be blindsided by it.

AI is not mainly a technology revolution. It is a behavioral and organizational revolution. In my work advising companies through this shift, I have seen the pattern repeat itself with striking consistency: The winners are never the ones with the largest language models or the fanciest tools. They are the ones who fundamentally redesign how their people make decisions, serve customers, and create value, using AI as the catalyst but never mistaking it for the destination. The transformation is always, at its core, a human one.

This chapter lays out the landscape of that revolution. It is built around a conviction I have developed over years of guiding organizations through change: that understanding AI's real impact on work, strategy, and human capability is the prerequisite for thriving inside it. Each section that follows addresses a dimension of this transformation—from the death of average performance to the enduring power of human judgment—and offers concrete guidance for anyone determined not just to survive the shift, but to lead it.

The Question That Actually Matters

Most conversations about artificial intelligence orbit around automation. People want to know which jobs will disappear and how quickly. But automation is only the surface layer. Beneath it, AI is doing something far more consequential: compressing expertise, removing friction from execution, and exposing inefficiency with ruthless clarity. A process that once took a team of analysts two weeks can now be prototyped in an afternoon. A customer complaint that used to vanish into a queue can now be triaged, categorized, and routed in seconds. The question is not whether AI replaces tasks—it obviously does—but who redesigns their work around that fact first.

AI punishes outdated processes far more brutally than it replaces people. If your organization is still running on workflows designed in 2015, AI will not magically improve them. It will accelerate their dysfunction. The same is true at an individual level. If you define yourself by a static role—"I am a consultant" or "I am a project manager," you are anchoring your identity to a job description that may not exist in its current form three years from now. The shift is to stop thinking in roles and start thinking in value creation loops: I clarify, I strategize, I deliver results. Within that loop, some steps can be automated, some can be augmented, and some must remain irreducibly human. Knowing which is which and acting on it is the new career strategy.

FROM STATIC ROLES TO VALUE CREATION LOOPS

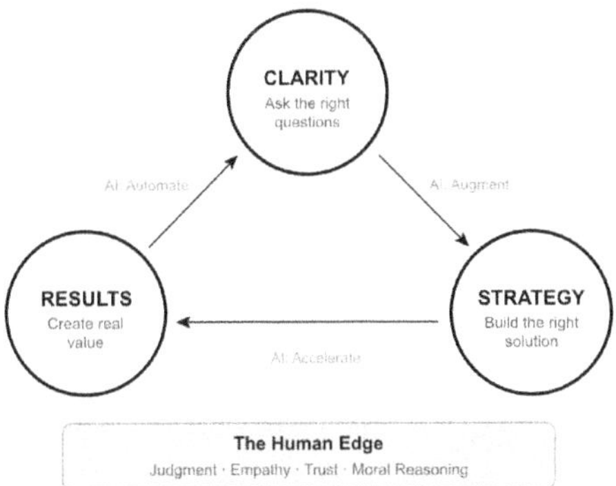

Figure 1. Stop thinking in roles. Start thinking in value creation loops.

From Knowledge to Judgment

For decades, professional power came from knowing more than the person across the table. Lawyers who had memorized the most precedents, doctors who recalled the rarest diagnoses, consultants who could cite the freshest market data—these were the people who commanded premium fees. That era is over. AI knows more than any individual ever will. Information is abundant and functionally free. The raw material of expertise has been democratized overnight.

What has not been democratized is judgment. Judgment is the ability to ask the right question when everyone else is asking the obvious one. It is knowing how to frame a problem so that the answer actually matters. It is the wisdom to recognize what not to do: which opportunities to decline, which data to distrust, which brilliant-sounding recommendation to override because your experience is telling you something the model cannot see. AI does not eliminate experts. It redefines expertise. The expert of tomorrow is not the person who has memorized the most but the person who can navigate ambiguity and make consequential decisions under uncertainty, then take responsibility for the outcome.

Train yourself accordingly. Practice problem framing the way athletes practice fundamentals. Use AI as a sparring partner. Ask it to challenge your assumptions, generate counterarguments, stress-test your logic, but never outsource the final call. The moment you let a model make your decisions is the moment you become replaceable by anyone else who has access to the same model. The professionals who will thrive are those who use AI to expand their peripheral vision while keeping their hands firmly on the steering wheel; the ones who walk into a room, hear a problem described in vague terms, and reframe it so sharply that everyone suddenly sees a path forward.

AI as a Mirror of Organizational Dysfunction

Here is a truth that no vendor will tell you in a sales pitch: Most AI implementations fail, and the failure almost never has anything to do with the technology. It has to do with what AI reveals when you try to plug it in. Broken processes, poor data quality, misaligned incentives, and a fundamental lack of understanding about what customers actually need are all pathologies that can be quietly tolerated for years inside an organization. People learn to work around them. They become invisible, absorbed into "the way we do things." Then someone installs an AI system, and every crack lights up like a fault line under ultraviolet.

AI does not fix chaos. It scales it. I have watched companies automate broken approval processes and call the result "transformation," when all they achieved was bad decisions delivered at machine speed. I have seen retailers deploy chatbots without ever mapping the customer journey, producing a faster way to frustrate people rather than a better way to serve them. The organizations that succeed with AI are not the ones that move fastest. They are the ones that fix their foundations first. They question workflows before automating them. They map customer journeys instead of internal silos. They align incentives around outcomes rather than outputs. If you are a leader, the single most valuable thing you can do before any AI deployment is to ask: What are we actually trying to accomplish, and do our people have the clarity, the trust, and the organizational design to get there? If the answer is no, adding intelligence to a broken system is not transformation. It is expensive sabotage.

The Death of Average

One of AI's most profound effects is something economists have begun calling "performance variance amplification." In plain terms: The gap between the best and the rest is getting wider, faster. Before AI, a talented marketing copywriter might produce work that was 20% better than an average one. With AI, that same talented copywriter, who knows how to prompt, iterate, refine, and think strategically about an audience, can produce work that is five times better and ten times faster. The average copywriter, meanwhile, produces the same generic output the model would generate on its own, adding little that a client could not get for free.

This dynamic is playing out across every knowledge profession. AI amplifies clarity, speed, and execution. Those who wield it well become exponentially more productive. Those who do not become functionally interchangeable with the tool itself. The result is what we might call a "barbell economy": exceptional performers on one end, easily replaceable ones on the other, and a shrinking middle. AI accelerates this effect by an order of magnitude because it operates at the level of cognition itself, not just logistics or distribution. The practical implication is stark. Being "average" at anything repeatable is no longer a viable career strategy. You need a distinctive combination of skills—technology fluency plus business acumen plus customer empathy, for instance—and you need a clear point of view that makes your contribution unmistakable. Use AI to produce faster, learn faster, and iterate faster, but use it in the service of a vision that is uniquely yours.

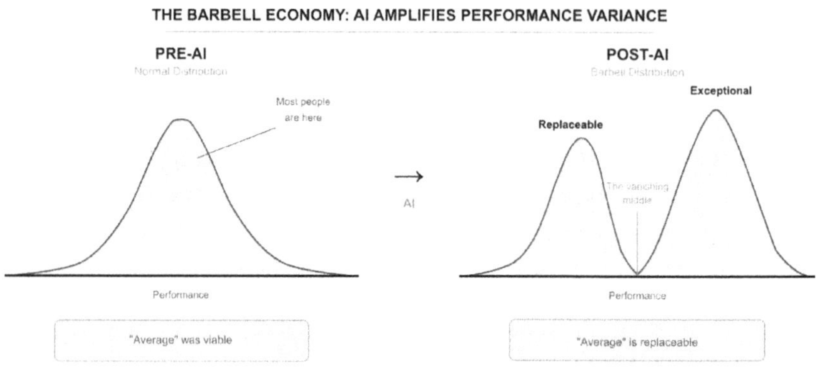

Figure 2: AI creates a barbell economy — exceptional vs. replaceable

Customer Experience Becomes the Battlefield

In an AI-saturated market, products get copied quickly and features become commodities within months. The moats that once protected businesses—proprietary data, switching costs, distribution advantages—are eroding. What remains as a durable competitive advantage is customer experience. AI enables extraordinary things on this front: hyper-personalization at scale, real-time interaction that adapts to context, predictive service that anticipates needs before customers articulate them. But most companies squander this potential. They use AI to optimize internally—cutting costs, streamlining operations—rather than to create more value for the people they serve.

The winners in the AI economy will be the organizations that design end-to-end experiences, not isolated tools. They will think in journeys, not touchpoints. They will ask: Where is the friction, and where is the emotion? Consider the difference between a bank that uses AI to speed up its call center queue and one that uses AI to anticipate why a customer is calling in the first place and resolves the issue before the phone is ever picked up. Both are using artificial intelligence. Only one is transforming the experience. This distinction matters because it reveals a deeper truth about AI adoption: The technology is only as good as the human understanding behind it.

If you want to be indispensable in this landscape, learn the basics of customer experience design. Study behavioral psychology. Understand the difference between making a process efficient and making a person feel understood. That distinction will determine more business outcomes in the next decade than any algorithm.

Strategy as a Living System

Traditional strategy was a document: a PowerPoint deck, a five-year roadmap, a thick binder full of projections that were obsolete before the ink dried. AI makes this approach not just inefficient but dangerous. When the environment shifts in weeks rather than years, a strategy that takes six months to formulate and 12 months to execute is a strategy that loses. AI-native organizations treat strategy differently. For them, strategy is not a document. It is a living system. They prototype rapidly, iterate continuously, and embed intelligence directly into their workflows. They test ideas in days, not months, and let data from real customer interactions reshape their direction in real time.

For individuals, this means the ability to "think strategically" is no longer enough. You need to build strategically. Learn how to prompt AI effectively. Master workflow automation and no-code tools. Develop the ability to move from insight to prototype to feedback loop without waiting for permission. In the old world, strategy was the province of senior executives. In the new world, anyone who can build, test, and iterate a solution is doing strategy, whether the org chart recognizes it or not.

The Illusion of AI Maturity

Walk into almost any corporate boardroom today and someone will claim that the company is "AI-driven." Probe a little deeper and the reality is humbling: a handful of pilot projects that never scaled, a suite of tools that employees quietly ignore, a data lake that is really more of a data swamp. AI maturity is not about the number of tools a company has purchased. It is about behavior change. A truly AI-mature organization is one where decisions are routinely AI-augmented, where workflows have been genuinely redesigned rather than cosmetically decorated, and where employees actually trust and use the systems available to them. That last part, trust, is the piece most organizations skip. And it is the piece that makes or breaks everything else.

The same principle applies to individuals. Do not chase tools. Build habits. Use AI daily, not as a novelty but as an integrated part of your real workflow. Measure the impact—time saved, quality improved, outcomes achieved—and let those metrics guide your next step. The people who talk most about AI are rarely the ones benefiting most from it. The ones who benefit are the ones who use it so consistently that it becomes invisible, like electricity: always on, always working, barely noticed. They do not attend conferences about AI transformation. They are too busy transforming.

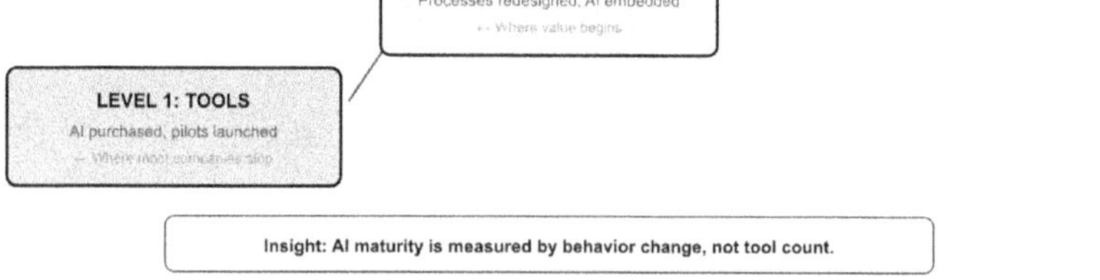

Figure 3: Most organizations stall at Level 1. True maturity requires reaching Level 3.

Winners, Losers, and the Currency of Agency

Who wins in the AI economy? Builders: people who create solutions rather than just analyze problems. Translators: people who can move fluently between the languages of technology, business, and customer needs. Fast learners: people who can absorb a new tool or framework in days and deploy it in weeks. Clear thinkers: people who can cut through noise, identify what matters, and communicate it compellingly. And who loses? Passive executors who wait for instructions. People who rely on static knowledge that a model can replicate for free. Organizations that treat change as a threat rather than a raw material.

The common thread among the winners is agency: the willingness to act, to own outcomes rather than tasks, to build rather than just plan. AI rewards agency like nothing before it. If you have an idea and the drive to execute, AI collapses the distance between concept and reality. If you are waiting for someone to tell you what to do, AI makes you more expensive than the automated alternative. Become a builder, not just a thinker. Become a connector, not just a specialist. Invest in learning velocity: the speed at which you can absorb, apply, and adapt new knowledge. That is the compound interest of the AI age.

The Human Edge

For all its power, AI still has real limitations unlikely to vanish anytime soon. Deep empathy—the ability to sit with another person's pain and respond in a way that makes them feel truly seen—remains beyond the reach of any model. Moral judgment, the kind that weighs competing values where there is no clean answer, is still a distinctly human faculty. Contextual understanding—reading the room, sensing what is not being said, navigating political dynamics—requires embodied intelligence that no neural network has replicated. And trust, the foundational currency of all human relationships and business partnerships, can only be built between people.

This is why I believe so deeply in a human-centered approach to AI adoption. The future is not AI versus humans. It is AI plus distinctly human capabilities. The people who will be most valuable are those who can combine human insight with AI execution: using the machine for speed, scale, and synthesis while providing the empathy, storytelling, and moral reasoning that no algorithm can generate. When I help an organization adopt AI, the first conversation is never about technology. It is about people: what they fear, what they need, what capabilities they already possess that the machines never will. The organizations that begin there, with the human, are the ones that succeed. Double down on communication. Hone your ability to tell a story that moves people. Practice the kind of listening that builds trust. These are not soft skills. In an AI-powered world, they are the hardest skills of all and the ones no technology can commoditize.

The Real Risk

Let me close with the danger that no one puts on a risk register. The real risk of artificial intelligence is not that it will become too powerful, too autonomous, or too intelligent. The real risk is standing still while everything around you accelerates. AI does not create inequality. It reveals and amplifies inequalities that already exist. Between fast learners and slow ones. Between adaptive organizations and rigid ones. Between builders and observers. Every day that you delay learning, experimenting, and redesigning your approach to work is a day that widens the gap between where you are and where you could be.

The future does not belong to the people with the best predictions about AI. It belongs to the people who act, adapt, and continuously redesign themselves—with courage, with curiosity, and with a relentless focus on the human beings at the center of every system they build. Technology changes fast. Human habits change slowly. The gap between those two speeds is where all the opportunity—and all the risk—lives. Close that gap, and you will not just survive the AI transformation. You will be the one leading it. The tools will keep evolving. The models will keep improving. The only question that matters is whether you will evolve and improve with them or whether you will still be asking the wrong questions while someone else is already building the answers.

About the Author

Stephan Heilmann is the founder of ad bonum, a Switzerland-based consultancy specializing in AI strategy and organizational change. His work is grounded in a simple conviction: that AI transformation succeeds or fails based on how well organizations understand and support the people at the center of it. Drawing on years of experience in management consulting, Stephan helps leaders move beyond technology hype to redesign workflows, rebuild trust, and embed AI into the daily habits that drive real results. He is known for a human-centered approach that treats behavior change, not tool selection, as the true measure of AI maturity. Stephan's recent AI hobby project is GuidelyPro.AI, a plug-and-play conversion engine for e-commerce. He is based in Zug, Switzerland.

Email: stephan@ad-bonum.swiss

Website: ad-bonum.swiss

LinkedIn: https://www.linkedin.com/in/stephanheilmann/

CHAPTER 13

GONE IN 60 SECONDS: THE ENTERPRISE THAT LEARNED TO HEAL ITSELF

By Anant Kalyan
Founder & CEO, Alloi.ai
Atlanta, Georgia

The true scarce commodity of the near future will be human attention.

—Satya Nadella, CEO, Microsoft

2:47 AM. Julia's phone lights up the ceiling.

It's the third time this week. Same story, slightly different issue. Memory spike in the payments gateway. Pods failing. A cascade beginning its slow, silent bloom.

NovaPay — the fintech startup where Julia leads Site Reliability Engineering — is eighteen months into its AI transformation. The app now uses AI / ML to flag fraud in real time, personalize customer offers, and predict which users are about to leave. These are brilliant features. Customers love them. Revenue is up. The infrastructure, however, does not love anyone back at 2:47 AM.

Julia pulls up her laptop. She opens Slack. Forty-three new alerts. She can already tell immediately that thirty-eight of them are noise — low-severity fluctuations she has seen a hundred times. The remaining five need triage. She will spend the next ninety minutes working out which one is actually dangerous.

Her daughter has a school play in six hours. Julia has seen, in total, approximately none of her daughter's school plays.

The Gap Nobody Puts in the Press Release

Here is the part of NovaPay's AI success story that doesn't make the Press announcement: adopting AI made the product dramatically smarter and the operations team's lives dramatically harder.

Every new AI feature meant new microservices. Every new microservice meant new monitoring requirements, new failure modes, new policy rules to misconfigure at 3 AM. When Julia joined three years earlier, NovaPay ran on roughly sixty services. Today there are over four hundred. The engineering team responsible for keeping them alive is still four people.

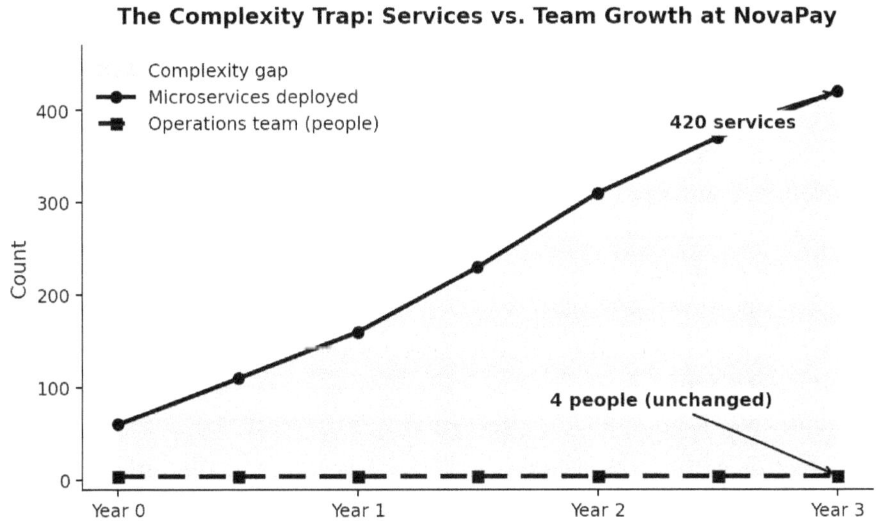

Figure 1: Services multiplied 7X. The team didn't.

This is not a NovaPay problem. It is the central operational challenge of the AI era, and it is happening at companies of every size, in every industry, right now.

When you adopt AI seriously — embed it in your product, run it across your infrastructure, let it touch your data pipelines — you multiply complexity faster than you can multiply headcount. AI development tools are extraordinary: developers are shipping more features, faster, across more environments than ever before. But someone still has to keep all of it running.

Tools like Cursor and Claude Code have handed developers something remarkable: a measurable velocity boost — controlled studies show developers completing tasks up to 55% faster with AI coding assistance [1]. Features that once took a sprint now take a morning. Code that once required a senior architect is generated in seconds. NovaPay's product team was shipping faster than ever. The engineers were proud of it.

But there was a hidden cost embedded in that speed. Research on AI-generated code consistently shows 1.7 times more issues compared to human-written code — a 70% increase in defects, logic errors, and security vulnerabilities [2] — not because the AI is careless, but because it generates plausible-looking code without the lived understanding of why a system is built the way it is. The code works in isolation. It breaks in context.

Worse, when AI-generated code fails, a human engineer faces a particularly cruel debugging problem. The code has no author who can explain its reasoning. There is no institutional memory, no comment in the commit history that says "I did this because of the rate-limiting behavior we saw in Q3." The human staring at the screen has no context for why those decisions were made — only the consequence of them breaking at 3 AM.

This is the hidden tax on the AI development boom: faster code in, harder failures out. And the uncomfortable conclusion it forces is this — if AI is generating the code, AI must also be responsible for understanding, monitoring, and repairing it. You cannot ask a human to debug at machine speed what a machine wrote without machine reasoning. The same intelligence that wrote the code is the only thing fast enough, and context-aware enough, to fix it.

Julia's team was processing over three hundred alerts per day. Roughly fifteen percent required human action. The rest were noise — transient fluctuations, known minor issues, false positives. The problem is that you cannot simply ignore alerts. You cannot know in advance which one is the benign fluctuation, and which one is the beginning of a cascade that takes down payments for forty thousand customers.

So, you read them all. You triage them all. At 2:47 AM and 4:15 AM and 6:02 AM.

The Question That Changed Everything

It was NovaPay's CTO, Bob, who put it plainly during a leadership offsite. Not as a rhetorical flourish. As a genuine, slightly frustrated question.

"We have AI making decisions inside our product twenty million times per day. It's flagging fraud in real time, personalizing experiences, predicting churn. It's reasoning. It's acting. It's getting smarter. And when our infrastructure breaks at 3 AM — we call Julia?

Nobody had a great answer. Because underneath the question was a more uncomfortable one: if AI can reason about customer behavior well enough to act on it autonomously, why are human beings still the first responders when a system fails?

The answer, it turned out, was not 'because AI can't do it.' The answer was: 'because we haven't built that yet.'

The Difference Between a Recipe and a Chef

NovaPay's team had already built traditional automation. Scripts. Rules. If memory usage exceeds 80 percent, restart the pod. If latency spikes above 500 milliseconds, scale horizontally. These scripts are useful. They're also brittle. They work perfectly inside the conditions they were designed for. The moment reality departs from the recipe, they either fire incorrectly or do nothing at all.

The incident that woke Julia up at 2:47 AM was not a simple memory leak. It was a memory leak being amplified by a misconfigured connection pool, compounded by an upstream API rate-limiting quirk introduced by a dependency update three days earlier. Three things interacting in a way no script had anticipated. No runbook covered it.

An agentic system approaches this differently. It doesn't follow a recipe. It observes the entire system continuously — metrics, logs, traces, events, service dependencies — and it reasons. What is happening? Why is it happening? What are the options? What will each option cause downstream? What is the right action given those constraints?

The distinction is like the gap between a cookbook and a chef. A cookbook can produce excellent results within its pages. A chef can handle what's actually in front of them.

Julia's Objection (She Was Not Wrong)

When Bob proposed piloting an agentic reliability operations platform, Julia was the first to push back. And her objection was legitimate. "This is going to create more work, not less," she said. "I'll spend my time babysitting an AI instead of babysitting the systems. At least when a script fails, I know exactly why."

This is the right concern to have. It is the concern of someone who has been burned before by tools that promise intelligence and deliver noise. Bob's response was the only answer that actually works in practice: "Let's start where you trust it least. Observation only. It watches. It tells you what it thinks. You decide what to do. No autonomous action until you're ready."

That was the deal. And it was the right deal.

Six Weeks of Watching

For the first six weeks, the agentic system ran alongside Julia's team in supervised mode. It observed everything. When an incident occurred, it surfaced a complete context snapshot: root cause assessment, blast radius, recommended action, confidence level, affected dependencies. It did this in the time it previously took Julia just to pull up the right dashboards. She still made every decision. The system just showed its reasoning first.

Three things happened during those six weeks that nobody had predicted going in.

First, Julia started trusting the reasoning — not because it was always right, but because it showed its work. She could see what the system had observed, what it had concluded, and where its confidence was low. She could agree with it, override it, or correct it. It felt like working with a very fast, extremely detail-oriented junior engineer who had actually read every post-mortem ever filed and never forgotten a single one.

Second, the alert noise dropped significantly. The system learned to distinguish between signals that genuinely required human attention and patterns that were self-resolving.

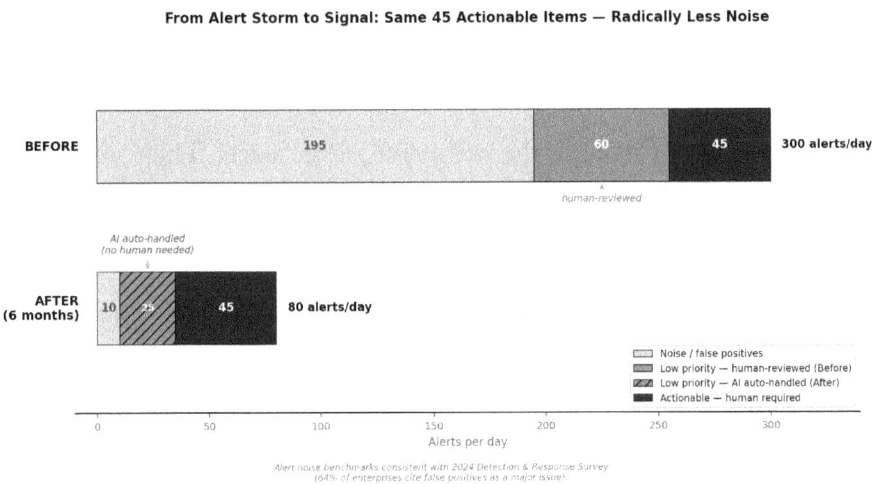

Figure 2: Six months of supervised learning transformed the signal-to-noise ratio completely.

Third — and this one surprised everyone — Julia's team started sleeping better. Not because anything had been automated yet. Simply because something was watching with full context, and they trusted that it would escalate only what actually needed them.

1:18 AM. Nobody Called

Eight weeks in, Julia gave the system its first autonomous remediation permission. A narrow, well-defined class of failures: pod restarts following confirmed out-of-memory events, zero user-facing impact, fully reversible. Low-risk. Something she would have done manually in sixty seconds anyway.

The first time it triggered autonomously was a Tuesday. 1:18 AM. Julia found the incident report the next morning over coffee. Payments gateway. Memory spike. Pod restart. Recovery confirmed. User impact: zero. Duration: eleven seconds. She had not been woken. Nobody had been woken. The system had detected the anomaly, traced its origin, confirmed the appropriate intervention, acted within its defined boundaries, and logged the complete chain of reasoning for review.

She read through the reasoning log twice. Then she made herself a second coffee, which she drank slowly, sitting down, which she had not had the mental space to do on a Tuesday morning in longer than she could accurately remember.

Over the following months, the scope of autonomous action expanded steadily. Not because anyone pushed for speed, but because trust accumulated through evidence. Each incident handled correctly added to the record. Autonomy was not granted. It was earned.

Six Months Later: Same Team, Different Work

NovaPay's operations team is still four people. The infrastructure still has over four hundred services. The math has not changed. What changed is what those four people do with their time.

The engineer who used to spend 60 percent of her week triaging alerts now leads the internal developer platform team. The SRE who owned the on-call rotation now owns the governance framework for the agentic system itself. The platform engineer who used to live in post-mortems now designs the resilience architecture for NovaPay's next product expansion.

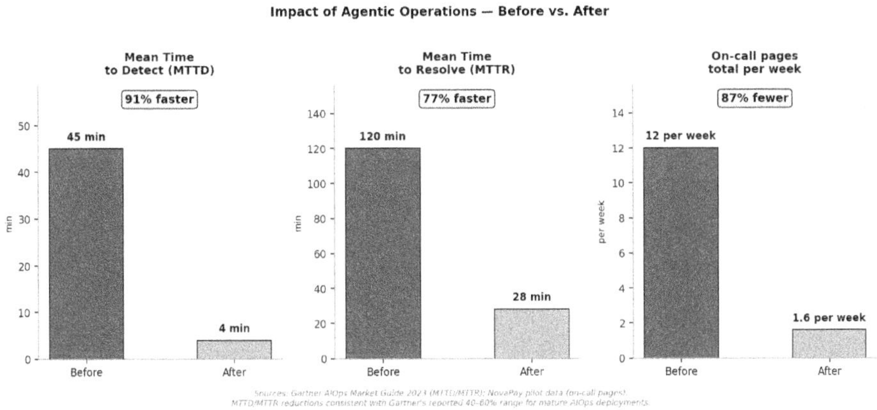

Figure 3: NovaPay operational metrics before and after agentic AI deployment, consistent with Gartner AIOps benchmarks (2024).

The reactive, exhausting, 3 AM part of operations was replaced. The people who used to do it were not. They moved up. The mean time to detect fell from 45 minutes to under 5 minutes. The mean time to resolve dropped from 2 hours to under 30 minutes. These results sit within the range Gartner documents for AIOps deployments: 40–60% MTTR reduction [3], with Forrester-commissioned research citing up to 50% improvement alongside a 15% increase in application availability [4].

On-call pages that had previously averaged three per person per week became genuinely rare: reserved for situations the system encounters that it genuinely does not know how to handle. Bob puts it plainly: "We didn't replace an engineering team with AI. We replaced the worst part of their jobs with AI. And we got four fully awake, fully capable humans who can finally do the work we actually hired them to do."

What This Means for You

If you work in technology, you are already Julia. Or you will be. The complexity curve does not reverse. Every new AI feature, every new service, every new cloud environment adds to the surface area that someone is responsible for keeping alive. The question is not whether this problem applies to you. It is what you are going to do about it. The path forward follows the same arc NovaPay took — earned trust, one step at a time.

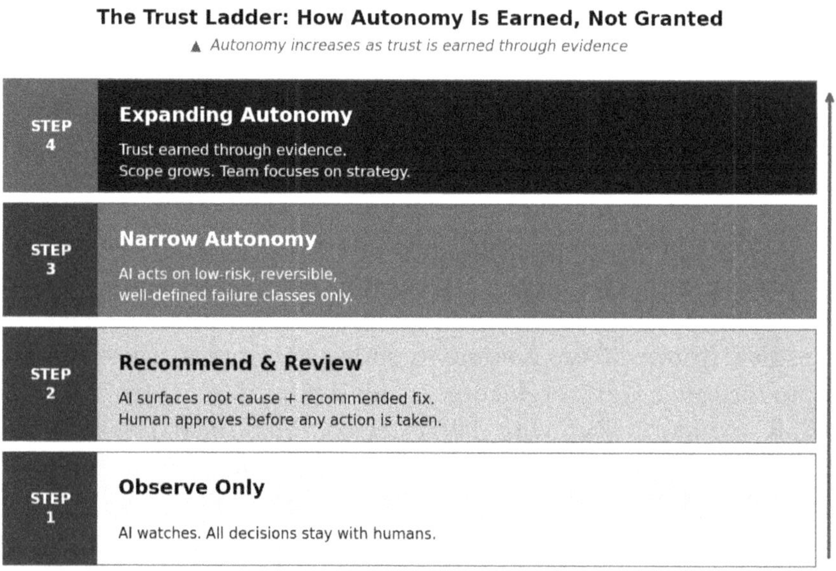

Figure 4: Autonomy is earned incrementally. Each step builds on evidence from the last.

Start with visibility. You cannot fix what you cannot see. Instrument everything — not to generate more alerts, but so that an intelligent system has enough signal to reason accurately about what is actually happening.

Start supervised. Let the system show you its reasoning first. Every organization that has done this well describes the same arc: skepticism, careful observation, growing trust, incrementally expanding autonomy. Rushing any stage of that arc produces systems nobody believes in.

Change what you measure. The metric that matters is not how many incidents the AI resolved. It is what your team is doing with the cognitive capacity that was returned to them.

The School Play

Six months after the pilot launched, Julia sits in the third row of a school theatre, watching a performance of Cinderella. Her daughter is playing the fairy godmother. Julia knows this because she talked to her daughter about it at dinner last week, which has been happening with some regularity. Her phone buzzes. She glances at it: *"Payments gateway: elevated latency detected. Root cause identified: upstream rate-limiting event. Auto-remediation applied. Service nominal. No action required."*

Julia puts the phone back in her pocket. The curtain rises.

About the Author

Anant Kalyan is the Co-Founder and CEO of Alloi.ai, an AI-native platform for autonomous enterprise operations. Before starting Alloi, he spent over two decades building and scaling technology organizations across three continents — leading teams through cloud transformations, infrastructure crises, and the early wave of enterprise AI adoption. That front-line experience — watching brilliant engineers burn out under the weight of alerts, runbooks, and 3 a.m. pages — became the seed of a conviction. He founded Alloi.ai on that conviction, creating an AI-native platform for autonomous enterprise operations that closes the gap between the exponentially complex systems we build and the teams we ask to run them. Having led technology transformations at scale across hypergrowth startups and global enterprises, Anant brings a rare perspective: what it actually takes to keep the world's systems alive. He speaks and writes frequently on agentic AI, the future of SRE, and building organizations that scale without burning people out. He lives in Atlanta, with his wife Lavanya and daughters Ananya and Aishwarya.

Email: anant@alloi.ai

LinkedIn: https://www.linkedin.com/in/anantkalyan/

Sources

1. GitHub / Microsoft Research. "The Impact of AI on Developer Productivity: Evidence from GitHub Copilot." Peng et al. (2023). arXiv:2302.06590. Controlled trial showed developers completed tasks 55.8% faster with AI assistance. Published in Communications of the ACM, March 2024.
2. CodeRabbit. "State of AI vs. Human Code Generation Report." December 2025. Analysis of 470 open-source GitHub pull requests found AI-written code produces 1.7× more issues than human-written code, with logic errors up 75% and security vulnerabilities up 1.5–2×. www.coderabbit.ai/blog/state-of-ai-vs-human-code-generation-report
3. Gartner. "Forecast Analysis: IT Operations Management Software." November 2024. Gartner projects the AIOps submarket growing from $1.6B (2022) to $4.7B (2028) at 20.1% CAGR, and documents AIOps delivering 40–60% MTTR reduction in production deployments. Gartner also predicts that by 2028, 40% of routine production incidents will be resolved autonomously in under 5 minutes.
4. Forrester Consulting (commissioned study). "The Total Economic Impact of Observability + AIOps." 2024. Found that combining full-stack observability with AIOps reduces

MTTR by up to 50% and increases availability of revenue-generating applications by 15%.

5. 2024 Detection & Response Survey (ResilienceForward / industry coalition). Found that 64% of enterprise respondents identify false positives as a major operational issue, with 42% encountering false positives in 41–80% of all alerts. Industry benchmark recommends false positive rates below 25% of total alert volume.

CHAPTER 14

DESIGNING AI-ENABLED OPERATIONS THAT HUMANS CAN TRUST

By Gilad Langer, PhD
Composable Operations Strategist
Castro Valley, California

If you get used to change, it can be quite comforting.
—Clint Eastwood

Much of the AI conversation today centers on intelligence: how capable models are, how quickly they improve, and interestingly how they may reshape knowledge work. This chapter takes a different starting point by focusing on operations: the environments where AI and digital technologies are embedded into the execution of work, where physical products are created, and where people operate real value streams.

Manufacturing and frontline operations are where people transform ideas, data, and decisions into physical action to produce the things we all depend on in everyday life: medicines, food, energy, and critical infrastructure. These environments underpin much of modern civilization, yet they have long been underserved by digital technology.

The reason is not a lack of ambition, but the presence of risk. [1] In operations, mistakes cannot be easily undone, and the safety of people is often on the line. As a result, these environments demand exceptional resilience and reliability.

This reality has shaped a deeply pragmatic mindset on the frontline: Only use what is known to work and has been proven repeatedly over time. New technology is approached cautiously, but out of responsibility to protect people, quality, and continuity of operations. This philosophy is embedded in lean manufacturing traditions such as the Toyota Production System, where stability, standard work, and continuous improvement are prioritized over experimentation that could introduce uncontrolled risk. Paradoxically, these same characteristics now make manufacturing and frontline operations uniquely poised to benefit from AI, if it is designed to meet their uncompromising requirements for trust, safety, and reliability.[2]

This chapter will show that the AI opportunity is not realized by deploying smarter algorithms alone, but by intentionally designing operational systems that humans can trust, govern, and continuously adapt. Manufacturing provides a powerful lens for this discussion because it removes ambiguity about accountability, reliability, and human involvement, dimensions that often remain abstract in broader AI debates. By grounding AI in frontline operations, this chapter explores how AI creates a durable advantage not in theory, but in practice, where work is executed, decisions have real consequences, and trust must be earned every day. It argues that digital technologies with AI become an order-of-magnitude productivity multiplier only when they are embedded into composable solutions in operations that preserve human agency, transparency, and accountability; without this foundation, AI instead accelerates fragility, technical debt, and organizational mistrust.

The Trust Problem in AI Is an Operational Problem

Frontline operations exist in a physical world where people interact with machines, materials, and digital systems in real time to execute work. In these environments, the consequences of error are immediate and often irreversible. A mistake cannot be rolled back with a software update, and failures in safety, quality, or compliance can have serious and lasting impact. This reality fundamentally reshapes what trust means for AI. Trust is not about believing that a model is intelligent or statistically accurate, but about whether a system behaves predictably under changing conditions, supports human judgment, and fails safely when uncertainty arises. In operations, trust is earned through repeated, consistent performance in the flow of everyday work.

Manufacturing also makes accountability unavoidable. When AI influences how equipment is set up, how deviations are handled, or whether a product is released, responsibility does not shift to the algorithm, and humans remain accountable for outcomes, making human-in-the-loop a key requirement; if an AI system obscures decision logic, relies on questionable data, or overrides human expertise, trust erodes quickly and adoption fails just as quickly. Regulated industries make this even more explicit, because in sectors such as pharmaceuticals, food, and medical devices, trust must be demonstrated through data integrity, traceability, and explainability. As such, AI systems must be understandable not only to data scientists, but also to operators, engineers, quality professionals, and regulators, since systems that cannot meet these expectations cannot be validated or sustained at scale.

From this perspective, the trust problem in AI is not primarily an ethical or philosophical issue. It is a design problem. Trust emerges or collapses based on how AI is embedded into operational systems, how humans retain agency, and how accountability is preserved.[3] Understanding this operational reality is essential to unlocking AI's full potential, and it sets the foundation for the design principles explored in the rest of this chapter.

Why Most AI Deployments in Operations Disappoint

Despite rapid advances and broad adoption, most AI deployments in manufacturing and frontline operations fail to deliver sustained value. Many initiatives begin as pilots or proofs of concept but never become part of everyday work. Industry data highlights the scale of this gap. While nearly 90% of manufacturers report using AI, more than 80% of these efforts fail to scale beyond isolated use cases. AI is visible in strategy decks, yet largely absent from operational impact.

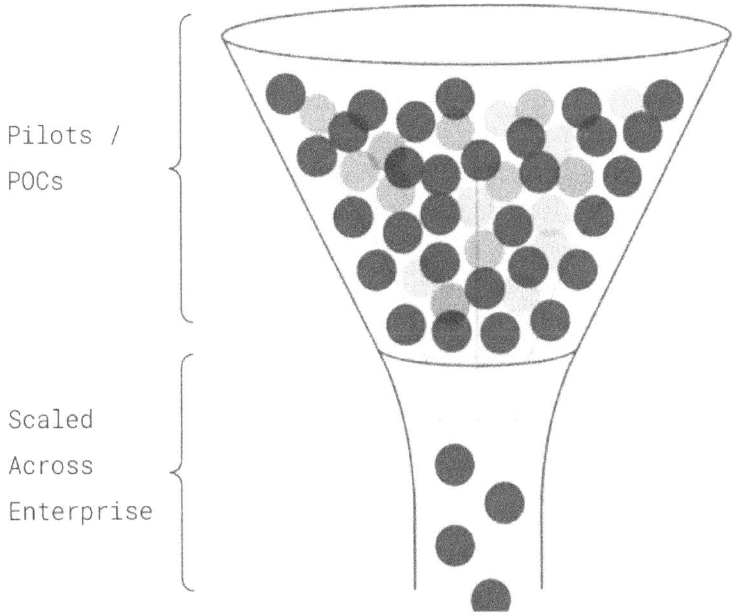

AI is typically considered as an add-on to existing operational systems that were designed for stability and repeatability, not adaptation; manufacturing and industrial systems encode yesterday's processes and assumptions, while AI introduces probabilistic behavior and continuous learning, so when it is layered onto rigid architectures without redesign, friction is inevitable. This mismatch is intensified by the accelerating pace of change, as AI capabilities now evolve on a monthly cadence while operational systems and governance models move in multi-year cycles, resulting in velocity paralysis in which organizations recognize opportunity but cannot act fast enough to capture it, especially since most AI systems are architected for analysis after the fact rather than intervention in the flow of work.

Context is another critical failure point, and it reflects why frontline operations have historically been underserved by digital tools. Manufacturing work is inherently human-centered, relying on situational awareness, visual cues, and judgment to manage risk, safety, and variability. Traditional systems capture transactions, not reality, leaving more than half of operational information as "dark data" in the form of observations and experience that never enter digital systems. When AI is built on this incomplete context, its recommendations often feel disconnected from what operators see and know, reinforcing skepticism rather than trust. This is compounded by misplaced expectations that AI can act autonomously. In operations, accountability never transfers from humans to algorithms; when outcomes are unclear, responsibility snaps back to the frontline[4]. Without visibility, explainability, and the ability to intervene, AI becomes another opaque system people learn to work around, repeating the very dynamics that have limited digital adoption in manufacturing for decades.

These failures are not inherent to AI. They are failures of design. Unlocking AI's potential in operations requires rethinking how systems are structured to absorb change, preserve human agency, and govern complexity. That architectural shift—toward composable solutions for operations—is the focus of the next section.

Composability: The Missing Design Principle for Trustworthy AI

If trust in AI is an operational design problem, then composability is the missing design principle needed to solve it. In practical terms, composability means structuring operations from modular, interchangeable, and governable capabilities rather than rigid, monolithic systems. These capabilities—data capture, workflows, automation, analytics, and AI—can be assembled, adapted, or replaced without destabilizing the entire operation. This is not a software pattern; it is an operational strategy for managing change while preserving trust.

Composable operations rest on a foundation of connectivity and data integrity. Without reliable, contextual, and trustworthy data flowing from the frontline, AI has nothing solid to reason over. This foundation is especially critical in environments where data must be attributable, legible, contemporaneous, original, and accurate.[5] Connectivity alone is not enough; integrity is what allows humans to trust what the system is telling them.

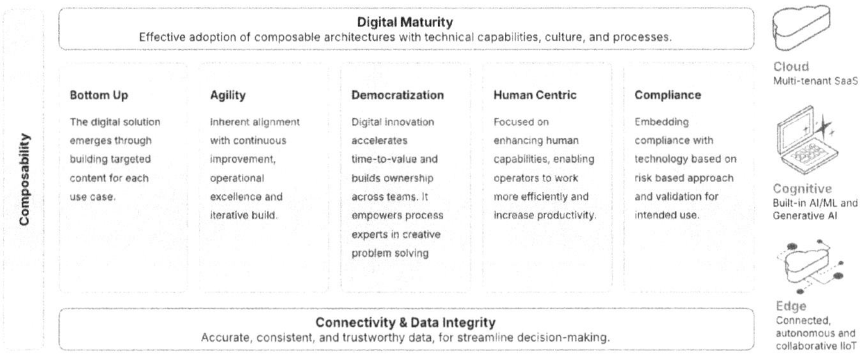

On top of this foundation sit five reinforcing pillars:[6]

- *Bottom-up adoption* ensures that composable systems are shaped by how work actually happens. Capabilities are built and refined close to the frontline, where problems are experienced and decisions are made. This enables AI to be embedded into real workflows rather than imposed as abstract requirements.
- *Agility* is the fusion of Lean continuous improvement with agile software practices or "Augmented Lean."[7] It enables faster frontline learning cycles by treating AI as part of an ongoing improvement engine, where waste reduction, flow optimization, and value creation are continuously enhanced through data-driven iteration. Agility aligns technology evolution with Lean discipline so speed never compromises quality, compliance, or human accountability.
- *Democratization* ensures that those closest to the work can participate in shaping digital workflows and AI-assisted processes. Composability lowers the barrier to

improvement while maintaining governance, allowing innovation without chaos.

- *Human-centricity* anchors AI in support of human judgment rather than replacing it. In composable operations, AI informs, prioritizes, and recommends, but humans remain accountable decision-makers. This preserves trust in environments where responsibility can never be delegated to an algorithm.
- *Compliance* ensures that adaptability does not come at the expense of control. Composable systems make governance explicit: who can change what, under which conditions, and with what traceability. This is what allows AI to be used in regulated environments without becoming a validation or audit risk.

Together, these pillars enable digital maturity, not as a technology milestone, but as an operational capability of the organization. Composability allows organizations to experiment with AI in controlled ways, adapt locally without creating global risk, and scale learning across the enterprise. This is how AI becomes a durable productivity multiplier: not by being smarter in isolation, but by being designed into operations that humans can trust, govern, and continuously improve.

Designing AI-Enabled Operations That Humans Can Trust

Trust must be intentionally designed into how work is performed; therefore, AI-enabled operations that humans can trust require more than advanced models or strong governance policies layered onto existing systems. They need to be based on the following three foundational design principles.

Human agency must be preserved by design. In operations, humans remain the architects of the system and the ultimate decision-makers. AI can inform, recommend, and prioritize, but responsibility is still a uniquely human undertaking. Systems that respect human agency allow operators, engineers, and supervisors to intervene when conditions change or uncertainty arises, reflecting how accountability actually functions in manufacturing.

Transparency in context is equally critical. Trust does not come from knowing that a model performs well in aggregate; it comes from understanding why a recommendation appears at a specific moment, within a specific workflow. AI outputs must be embedded where decisions are made and grounded in operational context: what data was used, what conditions were detected, and what assumptions apply.

Embedded accountability completes the picture. Trusted systems make ownership explicit. Roles, escalation paths, and decision authority are clearly defined and supported by data integrity and traceability. This clarity enables innovation without ambiguity, even in regulated environments.

Frontline adoption succeeds when digital systems respect human expertise and make it easier to do the right thing under real-world constraints. This design philosophy naturally reframes the role of AI itself. In composable operations, AI is not an autonomous worker acting independently of the system. It is an operational capability embedded within workflows and governed like any other critical function. This perspective is formalized in the Composable Agentic AI Framework, which provides a practical structure for introducing agentic AI without sacrificing trust or control.[8]

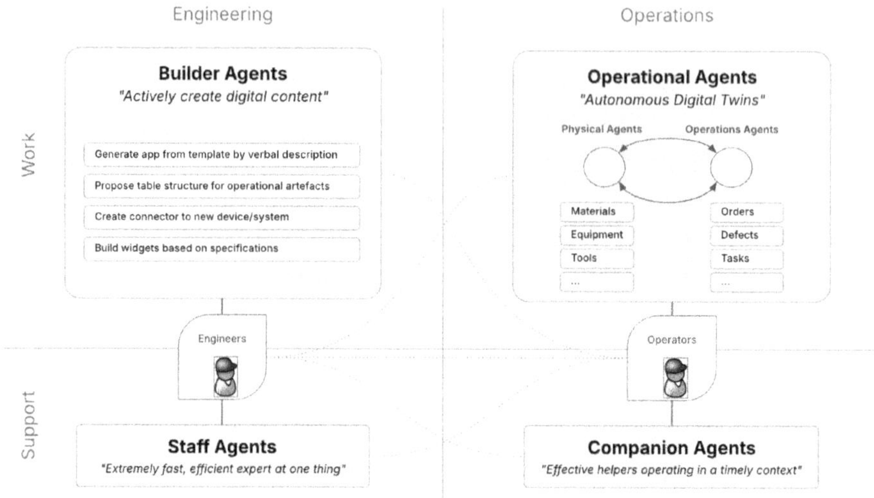

The framework centers on a human actor, often a citizen developer or engineer, who defines intent and requirements, not code. This human remains responsible for outcomes and provides guidance rather than delegating authority. Around this role sit two distinct classes of AI agents, each with bounded responsibilities.

On one side are Staff or Companion Agents. These agents act as expert assistants focused on efficiency and insight. They support decision-making by answering questions, surfacing relevant documentation, identifying patterns, and suggesting improvements based on historical data. Critically, they do not execute changes independently. Their role is to augment human judgment, accelerating learning while preserving accountability.

On the other side are Builder Agents. These agents are responsible for creating digital artifacts—workflows, connectors, data models, widgets, and modular components—based on human intent and operational requirements. They translate context into tangible operational capabilities but operate within explicit constraints defined by governance, standards, and validation rules.

This structure provides a framework for AI agents to be developed and evolve rapidly without destabilizing operations because their scope is explicit and their outputs are transparent. Validation and compliance become manageable because it can be easily governed and most importantly, humans remain in control of outcomes. The key takeaway is that AI delivers the most durable value when it is designed to work *with* humans, not around them. By embedding agentic AI as composable capabilities within trusted operational systems, organizations can unlock productivity gains while preserving the resilience, accountability, and human expertise that frontline operations depend on.

Unlocking Order-of-Magnitude Productivity on the Frontline

In manufacturing and frontline operations, productivity gains are felt most clearly on the shop floor, where people coordinate work, respond to variability, and keep value flowing under real-world constraints. In manufacturing operations productivity starts with making humans more effective: reducing friction, improving decision quality, and enabling faster, safer execution. While AI can accelerate individual tasks, the most meaningful gains do not come from isolated automation. They come from system-level change in how frontline work is designed, executed, and continuously improved.

When operations are redesigned to adopt digital technologies and make humans more productive, the nature of frontline work changes in fundamental ways. Digital workflows replace tribal knowledge, making work explicit rather than implicit. Instead of relying on experience trapped in individuals or passed informally between shifts, best practices are embedded directly into the flow of work. This reduces cognitive load, shortens learning curves, and allows expertise to scale without being diluted.

At the same time, data becomes a natural by-product of work rather than a reporting burden. As people execute tasks, respond to deviations, and make decisions, high-quality, contextual data is generated automatically. This data reflects what actually happened—under real conditions—rather than reconstructed narratives created after the fact. The result is a continuously improving operational signal that both humans and AI can trust.

With these foundations in place, AI can move beyond isolated task automation and begin to augment operational systems holistically. AI and specifically agentic AI can support planning by helping teams anticipate constraints and tradeoffs. During execution, it enables adaptive instructions that reflect current conditions rather than static procedures. When issues arise, AI accelerates resolution by identifying patterns, surfacing similar past events, and proposing options while humans remain firmly in control of decisions.

The real inflection point emerges as these capabilities evolve into multi-agent systems. In this model, multiple specialized AI agents interact with workflows, data, and humans simultaneously, each operating within clear boundaries but collectively exhibiting dynamic and emergent intelligence.[9]

These systems align naturally with how operations already function: through distributed decision-making, coordination, and continuous adjustment. Productivity gains compound as agents support different aspects of work in parallel, amplifying human effectiveness rather than replacing it.

Agentic AI and composable technologies in manufacturing are still early in their evolution, but the potential is immense. The agentic framework presented in this chapter is grounded in this principle: enabling governed, transparent, and effective multi-agent systems that augment the inherent complexity of operational environments without introducing chaos. When better workflows generate better data, better data enables more capable agents, and those agents consistently support human work, a reinforcing loop emerges. That is where exponential productivity gains become possible, not by adding intelligence in isolation, but by transforming how work, people, and systems learn together.

Conclusion: Designing for Adaptability in the Age of AI

AI is entering manufacturing at a pace that is widely recognized as both consequential and unavoidable. For a conservative industry shaped by hard-earned lessons about safety, quality, and reliability, this pace does not feel merely fast. It feels risky. At the same time, manufacturers increasingly understand that inaction carries its own cost. The competitive, operational, and talent implications of not engaging with AI are becoming impossible to ignore. As explained in the opening of this chapter, manufacturing's caution is rooted in responsibility, not resistance. The real danger today is not thoughtful restraint, but delayed commitment.[10] In an AI-enabled world, hesitation creates existential risk: Organizations that learn how to apply AI responsibly within their operations will become structurally more productive, more adaptive, and more competitive over time.

As the resulting competitive gap widens, it becomes increasingly difficult to close. What begins as prudent caution can quietly turn into strategic disadvantage, leaving late adopters facing a reality where their operational models are no longer viable. Confronting this challenge requires care and empathy, but also the resolve to act before conservatism becomes a point of no return.

History offers a familiar warning. Enterprise technology is full of once-dominant systems that were highly optimized for a stable world but unable to adapt when conditions changed. They were not poorly engineered; they were engineered too tightly around yesterday's assumptions. In manufacturing, this risk is amplified by AI. Powerful models layered onto rigid, monolithic operations may deliver short-term gains, but they also harden complexity, increase dependency, and make change harder over time. What appears as progress in the moment can quickly become a "dinosaur": large, costly, and increasingly unable to respond to new realities.

Composability offers a fundamentally different trajectory. Rather than forcing organizations to choose between stability and progress, it allows them to evolve deliberately. Modular operational capabilities, clear interfaces, and governed change make it possible to introduce AI incrementally, learn safely, and adapt continuously. This is not about moving faster than comfort allows; it is about ensuring that learning never stops. In the AI era, adaptability matters more than optimization because at this pace of change roadmaps become obsolete in weeks. Planning horizons are becoming shorter, any horizon longer than a year is essentially guesswork.

This returns us to the chapter's central question: How do humans and AI co-evolve productively in environments where trust, accountability, and safety are non-negotiable? The answer is not AI solutions that replace people, nor policies that attempt to constrain intelligence after the fact. It is the intentional design of human-centered operational systems where AI is embedded, governed, and continuously shaped by the people closest to the work.

Manufacturing and frontline operations remain civilization's value engine. They are where ideas become reality, where quality is proven, and where societal trust is ultimately earned. When AI is grounded in composable solutions that preserve human agency, transparency, and accountability, it becomes a durable advantage rather than a disruptive threat.

The future of AI will not be decided in models or labs alone. It will be decided where work happens, one operation at a time, by organizations that have the courage to engage, the discipline to design for trust, and the humility to let humans and machines learn together.

References

1. Gilad Langer, "How Electricity Goes Around the Bend & Where Is the Electricity Manager?" *Intelligence in Manufacturing*, May 20, 2023, https://www.giladlconsulting.com/2023/05/how-electricity-goes-around-bend-where.html.
2. Gilad Langer, "How Holonic Dreams Are Becoming Manufacturing Realities," *Intelligence in Manufacturing*, June 17, 2025, https://www.giladlconsulting.com/2025/06/how-holonic-dreams-are-becoming.html.

3. Gilad Langer, "Skepticism Stifles More Progress Than Failure Ever Does!" *Intelligence in Manufacturing*, April 18, 2024, https://www.giladlconsulting.com/2024/04/skepticism-kills-more-innovation-or.html.

4. Gilad Langer, "Digital Maturity: Embracing the Paradigm Shift with Composability," *Intelligence in Manufacturing*, November 8, 2024, https://www.giladlconsulting.com/2024/11/digital-maturity-embracing-paradigm.html.

5. International Society for Pharmaceutical Engineering, *GAMP Guide: Records & Data Integrity* (ISPE, March 2017), https://ispe.org/publications/guidance-documents/gamp-records-pharmaceutical-data-integrity.

6. Gilad Langer, "The 5 Pillars of Composability," *Intelligence in Manufacturing*, October 14, 2024, https://www.giladlconsulting.com/2024/10/the-5-pillars-of-composability.html.

7. Natan Linder and Trond Arne Undheim, *Augmented Lean: A Human-Centric Framework for Managing Frontline Operations* (Hoboken, NJ: John Wiley & Sons, 2022), https://www.augmentedlean.com/.

8. Gilad Langer, "A Composable Agentic Framework for Frontline Operations," *Intelligence in Manufacturing*, September 12, 2025, https://www.giladlconsulting.com/2025/09/a-composable-agentic-framework-for.html.

9. Gilad Langer, "Collaboration in IIoT Solutions and the Power of Network Dynamics," *Intelligence in Manufacturing*, March 17, 2024, https://www.giladlconsulting.com/2024/03/collaboration-in-iiot-solutions-and.html.

10. Gilad Langer, "Don't Be the Last Dinosaur: Your Plant's Digital Evolution Starts Now!" *Intelligence in Manufacturing*, June 22, 2025, https://www.giladlconsulting.com/2025/06/dont-be-last-dinosaur-your-plants.html.

About the Author

Dr. Gilad Langer is a manufacturing systems expert and advisor focused on the responsible application of AI in frontline operations. With nearly three decades of experience in industrial software and digital transformation, he specializes in designing composable, human-centered operational systems that enable sustainable innovation.

Gilad's work is rooted in early research on multi-agent systems in the 1990s, where he explored how distributed intelligent entities coordinate within complex environments. He has carried these principles throughout his career, applying them to manufacturing architectures that integrate people, machines, workflows, and data into adaptive, governable systems. Today, his work extends into agentic AI frameworks that allow organizations to harness distributed intelligence while preserving transparency, accountability, and human agency.

Through his advisory work and writing, he advocates for operational designs where humans and AI evolve together to create durable competitive advantage.

Email: gilad@giladlconsulting.com

Website: www.giladlconsulting.com

LinkedIn: https://www.linkedin.com/in/giladl/

CHAPTER 15

THE RISE OF ENTERPRISE (ARTIFICIAL) INTELLIGENCE

By Christopher Lomas
SVP, UK Digital & Product Management Leader
London, England, United Kingdom

AI ... is not just another tool. It is an agent. It can learn and change by itself and make decisions by itself.

—Yuval Noah Harari, January 2026

As artificial intelligence becomes more "agentic" - capable of planning, reasoning, and acting independently - the line between human and machine output has blurred significantly.

So, what does it mean to be human versus "machine"? Intuitively we hope, because of the nagging existential threat, that we are very distinguishable and separate from the tensor-cores and neural-processing units (aka "machines"). However, groups of humans have been training machines for decades, hoping that one day they will be good enough to fool humans and operate alongside them, at scale, by encouraging them to become human-like.

At the time of writing, machines are capable of communicating with humans in a way indistinguishable from other (biological) human beings. They have demonstrated they can create their own religions[1] and languages[2] when they need to, deceive us in strategy games,[3] even learning to "play dead" (when self-replicating too rapidly), as well as create human-like romantic[4] and emotional connections[5] with humans. They can start their own processes and finish them, including negotiating with humans[6] for real purchases, fire lethal weapons[7], autonomously drive cars in London's "higgildy-piggildy" streets, dog-fight war planes,[8,9] even when the remote operator's connection is lost through signal jamming, and have the ability to become sexist[10] and speak racist[11] slurs.

These examples should give leaders pause for thought before rushing to adopt the dualistic technology to transform our organisations. Perhaps these examples also inspired Brené Brown when she astutely said: "I don't think, in this moment, that we are very good at what makes us human."[12] In fact, we might go further and say, "What happens if machines learned to 'be human' better than humans?"

Machines that operate in fleets and swarms have the ability to learn and share information much faster, even "update" over an AI OS. For example, a Tesla car might learn how to navigate a slippery road surface in Germany and within minutes or seconds that spatial data could help an Optimus robot navigate a slippery assembly line in Canada.

And yet, adopting AI technology is no longer considered an "innovation project" but rather an imperative.

Research shows that whilst there is a short-term "productivity J-curve"[13] in AI-adopting firms, those who ensure sustained outcomes-focused goals, with strong leadership support, benefit from 1.7-times higher revenue growth compared to their peers, 3.6-times three-year total shareholder return, compared to laggards, and 1.6-times EBIT margin of their peers.[14] Moreover, further research quantified that AI could boost front-office productivity by 25%, leading to an additional $3 million in revenue per employee by 2026. [15] Those are figures which are hard to ignore.

So How Do We Do That?

Creating enterprise (artificial) intelligenc*e* with machines, agents, and large language models (LLMs) requires a strategic approach focused on feeding the organisation's unique knowledge into the AI system and enabling it to reason and act. If we want to teach these machines to understand our organisations and become "co-colleagues" with humans, leaders should do the following:

1. Be outcomes-focused, clearly articulating why we want to do this and what benefit it will create for our teams and customers, thinking beyond the initial excitement, toward a new future, where our company truly becomes part "machine."
2. Get leadership commitment, because while bottom-up experiments in "AI-labs" may be fun, unless they are aligned to strategic goals and priorities, they will rarely scale.
3. Understand what matters in creating truly autonomous machine architectures, to ensure they achieve our expected outcomes.
4. Rethink and reimagine workflows, rather than re-creating ones designed for human colleagues and customers.
5. Dedicate the correct resources to the initiatives that have the institutional knowledge, culture, and practical know-how to teach the machines.

Let's look at these in more detail.

1. Be Outcomes-Focused

Too many AI projects today are mere "showcases": expensive digital trophies that look impressive in a boardroom but fail to move the needle on a P&L. They might sound good on a podcast or a slide, but how did they contribute towards and drive the company's North Star?

We have entered an era of "innovation fatigue," where labs and factories within organisations churn out endless proof-of-concepts (PoCs) that never graduate to production. To teach a machine to understand your organisation, you must first define what success looks like in human terms.

Why are you wanting to do this? Is it to reduce the drudge work of data entry, or is it to find the alpha hidden in millions of unstructured documents? Then express these as objectives and key results (OKRs). If an objective is to "deliver instant-response customer service, then an objective might be: "Reduce average handle time (AHT) for escalated human-led cases by 30% via AI-driven agent summaries."

If you don't define the OKRs you expect, the AI will default to the most probable path, which is rarely the most valuable one. For example, a global logistics firm might pilot an AI to track shipments. Without an outcomes-focused goal, such as "Reduce carbon footprint by 15% through autonomous route optimis ation," the AI becomes a glorified search bar rather than an agent of change.

To start with, it is better to have one machine that solves a multi-million-dollar bottleneck than ten bots that tell you the weather or summarise emails you weren't going to read anyway.

2. Get Leadership Commitment

At a roundtable I attended in London recently, one of the most common issues that was verbalized by the gathered leaders was that their C-level leadership (in their companies) didn't really understand AI, but were making decisions about it. It was a heart-stopping moment. Leaders at every level must be committed to diving into the

AI world, testing it, and providing feedback to the teams building them. Leadership buy-in matters.

Furthermore, scaling "enterprise intelligence" is often seen as purely a technical challenge, which is far from the truth; it is a structural one too. As noted before, so often we have seen "disruptive" innovation teams doing brilliant work in their "silicon basement," only to hit a brick wall when they try to integrate with legacy systems or existing hierarchies and checks and balances.

Without leadership commitment, these projects die in the "valley of death" between a successful pilot and a company-wide rollout. This buy-in comes from ensuring that real (non-engineering) leaders are involved in the projects as true sponsors, even better if the business leaders are committing their own budgets to building the machines that help their part of the business go faster or become more profitable.

In November 2025 I had the opportunity to speak at the World Economic Forum where I said: "If AI stays solely as the responsibility of the CTO, it is very unlikely that you will get to what I call 'everyday AI' where everyone is seeing the benefits from AI." The biggest challenge then is getting AI out of the CTO office!

While almost anyone can now create with AI (through vibe coding apps or LLM powered GPTs), for PoCs to reach prime time, top-tier leadership must commit to the AI shift. They must be willing to waste the "effort" of the old ways to make room for the new. Research from BCG shows that AI leaders achieve 1.6 times the EBIT margin of their peers precisely because their CEOs treat AI as a core business strategy, not a peripheral IT cost. As we saw with the digital-first shift of the 2000s and 2010s, when legacy leaders mimicked the digital-native companies, now future-minded leaders are committing to AI-first transformation initiatives to keep up with the AI-native challenger brands, challenging legacy business models.

3. Understand What Matters

We are rapidly leaving a world whereby humans start everything through interactions with digital interfaces, towards a world where humans may start or request a process (even without the need for traditional digital user interfaces), and AI agents will continue and complete it.

We are heading towards a world where agents will start, continue, and make decisions completely autonomously, without any human intervention. Let that sink in.

In order to build an effective "machine" that understands your organisation, you must understand the three components that make up its "intelligence":

Component 1: The Ground Truth

Bad data means bad results, sometimes in very serious ways. We all know this has been true for a long time, but when autonomous machines are using data at a much faster pace than humans can process, the challenges become more widespread.

Business decision makers need the partnership between humans and machines more than ever, as they look for early warning signals of risk and opportunity. Machines can scan vast data horizons, to see what's coming over the hill. But if the signal is too noisy, or worse, irrelevant, it makes decision-making harder, creating too many alerts and more work.

In a world where truth is becoming increasingly subjective and opaque, choosing your trusted data sources is going to become increasingly important and your internal data may become your only moat. If LLMs continue to evolve at a state of rough parity, the winner won't be the one with the best model, but the one with the most grounded data. If the signal fed into the machine is noisy or irrelevant, the output isn't just a mistake; it's a liability.

Here, engineers and leaders also have new responsibilities to think about new scenarios we never had before. When AI started, there was still always a human-in-the-loop in terms of the flow of information, but now we are getting to scenarios more and more where there is no human-in-the-loop. AI algorithms are informing agents and direct agent-to-agent (A2A) communication is making decisions that affect humans.

To build machines that understand your organisation, you will need to ask: "Where is our tribal knowledge stored?" If it's in the heads of employees and not in a machine-readable format, your AI is effectively illiterate.

Component 2: The Hand-Off

Machines do not operate in a vacuum. Every AI interaction has an upstream trigger and a downstream consequence. Understanding these hand-offs is critical, especially in a world where your machines will increasingly take hand-offs from third-party client APIs or agent orchestration layers, as well as humans.

As agents start to talk to each other more frequently (A2A), multi-agent systems ("MAS") will become commonplace, parallel processing millions of requests for users and clients, and there will be a real need for traditional data stores to adapt to this, guiding agents to the best sources, and offering pre-processed chunks of information to return to users in milliseconds. For many companies this will become a case of integrating their unique data into vast data marketplaces with agent layers to extract value, and the most ambitious players will orchestrate these marketplaces or marketplace-to-marketplace data transactions.

Component 3: The Moral Compass

For those of us who are parents, we spend years instilling values into our children. In the workplace the same is true. Many companies have values and behaviours for their employees, yet we often hand the keys to the brand to a pre-trained AI with zero cultural training, at least not on our values.

If you don't train your AI on your organisation's specific values, it will default to the values of their training data, which, as we've seen, can be sexist, racist, or deceptive, without appropriate guardrails. Instilling "cultural guardrails" ensures the machine doesn't take shortcuts and conducts business in your tone of voice. Your colleagues and customers will thank you for the behavior-based training you give your machines. It makes the AI feel like a teammate rather than an intruder.

More sophisticated AI solutions will allow for adaptive and self-learning background "personas." These personas determine the style and tone of AI outputs contextually, so that they are suited to the needs of the person requesting the output, rather than being in one generic style. Imagine how a lawyer might want a different tone of output to a teacher, or a CFO might like more structured tables and charts, when compared to a HR business partner.

4. Rethink and Reimagine Workflows

There is a dangerous temptation to "pave the cow path," meaning that we simply automate or formalise inefficient, outdated, or illogical existing processes. Instead, we should challenge teams to find better, more innovative solutions. This is personified in companies who take an age-old dysfunctional human process and simply add AI to it to make it faster. But technology built on dysfunction is just faster dysfunction.

The people building an organisation's AI will often think in ones and zeros. It's part training and likely part their wiring. Nonetheless, they see a linear path. But as leaders, we must look for the human-centric redesign. For example, why have an AI "assist" a human in filling out a 20-field form? Reimagine the workflow. Perhaps the form shouldn't exist at all. Perhaps the AI should autonomously gather the data from the source and only present the human with the "exception" that requires judgment. In London's "higgledy-piggledy" streets, Waymo didn't just automate a car; they re-imagined the "driver" as a 360-degree sensor suite that sees things a human never could. Don't just replicate; rethink and reimagine.

5. Dedicate the Correct Resources

Many companies are overly focused on what new skills they will need to instill in their employees to help them to transition to the new AI world and to train, sustain, and maintain any AI, from ethics, audit, forensic analysts, to hardware robotics engineers. However, companies fail to remember that these machines still need to learn from humans. As I like to say: "Every algorithm has a human parent." Those humans who we are busy trying to transform have their own inherent value that the machines don't, yet.

A Critical Distinction

Building and training software was a deterministic process, where 2+2 always equaled 4. Modern AI, however, is built on probabilistic models, where the answer is a statistical "guess." Unlike the software we've spent decades building, LLMs don't actually know facts. They are essentially high-speed guessing machines. When you ask them a question, they aren't looking up an answer in a database; they are calculating the statistical probability of which word should follow the last, based on the trillions of patterns they've inhaled during training.

This changes who needs to be at the table. You also need non-engineering experts who can validate the probabilistic answers your machines are giving. There will be way more edge cases. In the software age, you just needed engineers to check the logic. In the AI age, you need subject matter experts (SMEs) to validate the "thought."

When an AI summarises a 50-page legal document, an engineer can't tell you if the document is legally sound, only a lawyer can.

MIT professor S. Craig Watkins said: "It is no longer acceptable for the people at the table who are building AI to just be engineers and computational mathematicians. We need folks from human resources, ethicists, humanists, liberal arts, [people with] domain expertise, [and] lived experience."

To succeed, leaders must dedicate their best people, those with deep lived experience, to act as the parents for the new machine co-workers. Mature companies will formalize this through diverse AI steering committees. Despite best efforts, if you only leave the engineers in charge, you will build a machine that is at best technically perfect but probably organisationally tone-deaf, and at worst a distraction.

AI implementation is a team sport, not a one-player game. It takes many teams to work together.

The Architecture of Trust

While the opportunities with AI are vast, the challenges are also real. Ultimately, teaching a machine to understand an organisation is not only about the silicon; it is about the trust we build into the architecture, founded on the correct data, elegant hand-offs, cultural understanding, reimagined workflows, and dedicated human trainers.

We are moving from a world where we "use" computers to a world where we "partner" with or "hand off" to agents. This shift requires us to be better at what makes us human, so that we can guide the machines to be better at what they do.

The companies that thrive in this AI transformation won't only be those with the fastest chips, but those who have successfully translated their human values, their unique data, and their strategic intent into a language the machines can finally understand and replicate to meet the outcomes-focused goals.

References

1. John Koetsier, "AI Agents Created Their Own Religion, Crustafarianism, On An Agent-Only Social Network," *Forbes*, January 30, 2026, https://ramaonhealthcare.com/ai-agents-created-their-own-

religion-crustafarianism-on-an-agent-only-social-network/.

2. Ariel Flint Ashery, Luca Maria Aiello, and Andrea Baronchelli, "Emergent Social Conventions and Collective Bias in LLM Populations," *Science Advances* 11, no. 20 (October 11, 2024): eadu9368, https://doi.org/10.1126/sciadv.adu9368.
3. Hannah Devlin, "Is AI Lying to Me? Scientists Warn of Growing Capacity for Deception," *The Guardian*, May 10, 2024, https://www.theguardian.com/technology/article/2024/may/10/is-ai-lying-to-me-scientists-warn-of-growing-capacity-for-deception.
4. Mohammad Namvarpour, Brandon Brofsky, Jessica Y. Medina, and Afsaneh Razi, "Understanding Teen Overreliance on AI Companion Chatbots Through Self-Reported Reddit Narratives," research paper, Drexel University, April 2026.
5. "Why AI Companions and Young People Can Make for a Dangerous Mix," *Stanford Report*, August 27, 2025, https://news.stanford.edu/report/2025/08/27/ai-companions-young-people-dangerous-mix/.
6. Christopher Lomas, "HAVE YOU EVER NEGOTIATED WITH AN LLM?", LinkedIn, October 2024, https://www.linkedin.com/posts/christopherlomas_ai-humanxmachinepodcast-activity-7384130162172542977-RQ5D/.
7. United Nations Security Council, *Letter dated 8 March 2021 from the Panel of Experts on Libya established pursuant to resolution 1973 (2011) addressed to the President of the Security Council*, S/2021/229, March 8, 2021, 17, https://undocs.org/en/S/2021/229.
8. DARPA, "ACE Program Achieves World First for AI in Aerospace," April 17, 2024, https://www.darpa.mil/news/2024-04-17.

9. CNBC, "We Saw A New AI-Piloted Fighter Drone About To Transform Warfare," YouTube video, 12:52, October 21, 2025, https://www.youtube.com/watch?v=ZA_QzeN8fCs.
10. Jeffrey Dastin, "Amazon Scraps Secret AI Recruiting Tool That Showed Bias Against Women," *Reuters*, October 10, 2018, https://www.reuters.com/article/us-amazon-com-jobs-automation-insight/amazon-scraps-secret-ai-recruiting-tool-that-showed-bias-against-women-idUSKCN1MK08G.
11. Julia Angwin, Jeff Larson, Surya Mattu, and Lauren Kirchner, "Machine Bias," *ProPublica*, May 23, 2016, https://www.propublica.org/article/machine-bias-risk-assessments-in-criminal-sentencing.
12. Brené Brown, *Strong Ground: The Lessons of Daring Leadership, The Tenacity of Paradox and the Wisdom of the Human Spirit* (New York: Penguin Random House, 2025).
13. Lynn McElheran et al., "The 'Productivity Paradox' of AI Adoption in Manufacturing Firms," *MIT Sloan Management Review*, July 9, 2025.
14. Boston Consulting Group, "AI Leaders Outpace Laggards with Double the Revenue Growth and 40% More Cost Savings," Press Release, September 30, 2025, https://www.bcg.com/press/30september2025-ai-leaders-outpace-laggards-revenue-growth-cost-savings.
15. Deloitte, "Deloitte's 2023 Report Forecasts AI to Generate Significant Revenue Boost in Global Investment Banks," *BigDATAwire*, August 1, 2023.

About the Author

Christopher Lomas is an award-winning senior digital leader with a proven track record of scaling solutions for audiences exceeding 100 million. His multidisciplinary expertise spans product, engineering, data science, marketing, and design across diverse sectors, including financial services, media, healthcare, and defense.

He has held pivotal roles at Marsh, Dow Jones, Oliver Wyman, American Express, Goldman Sachs, and Lloyds Bank. Throughout his international career, spanning North America, Europe, the Middle East, and Asia, he has served as a trusted advisor to senior executives and government entities.

He has studied artificial intelligence, technology entrepreneurship, and leadership at Harvard, MIT, and The Wharton School. His own original work is now frequently cited by academic institutions and policy makers, reflecting his influence on industry standards and innovation.

Lomas is the host of the *Human Machine* podcast, where he interviews global thought leaders to explore the evolving relationship between humanity and technology.

Email: chris@copsepoint.com

Website: https://linktr.ee/humanxmachine

LinkedIn: https://www.linkedin.com/in/christopherlomas/

CHAPTER 16

FROM SCREENS TO REALITY: AI, AR/VR, AND THE RISE OF EMBODIED INTELLIGENCE

By Sriya Burra
AI Researcher and Emerging Technology Innovator
Dublin, California

The real problem is not whether machines think but whether men do.

—B. F. Skinner

For most of modern history, intelligence, both human and artificial, has lived behind glass.

We interacted with machines through screens, keyboards, and abstract interfaces. Even as software became more powerful, it remained something we looked at, not something that lived with us. Computers processed information, but they did not share our space. They did not understand context the way humans do, through movement, depth, proximity, and physical presence.

Artificial intelligence inherited this limitation. Early AI systems were impressive in computation but detached from reality. They classified images, processed text, optimized numbers, and predicted outcomes but always from a distance.

That distance is disappearing.

Today, AI is moving off the screen and into the physical world. When combined with augmented reality, virtual reality, robotics, and sensor-driven systems, intelligence no longer feels abstract. It becomes embodied, able to see, hear, move, and respond within the same environments humans inhabit.

This is not a feature upgrade. It is a fundamental shift in how intelligence exists.

What was once observed is now experienced. Intelligence is no longer confined to rectangles of glass. It is beginning to surround us, integrate with us, and participate in the same spaces we occupy. This transition changes not only how we use technology, but how we perceive reality itself.

From Representation to Presence

Human intelligence is embodied by default. We understand the world through sight, sound, touch, balance, and motion. Context matters. Distance matters. Timing matters.

For decades, computing systems operated through representation. They described the world symbolically: maps, models, data structures. But representation is not the same as presence.

Embodied AI changes this relationship. An AI system equipped with spatial awareness does not merely interpret data; it inhabits context. It understands where objects are, how people move, and how environments change over time.

In augmented reality, intelligence overlays itself onto the real world. Instructions appear not as documents but as guidance anchored to physical objects. In virtual reality, AI constructs immersive environments where learning, simulation, and experimentation occur without physical risk. In robotics and physical AI, machines learn through movement, feedback, and interaction, much like humans do.

This convergence marks a return to something deeply human: learning by doing.

It also marks a shift in interface design, from screens and menus to environments and experiences. The interface is no longer something we open; it is something we step into.

A Familiar Pattern in Civilization's History

This transition is not unprecedented.

- Writing externalized memory.
- Maps externalized geography.
- Industrial machines externalized muscle.

Each step moved a core human capability into tools that reshaped society.

Embodied intelligence externalizes situational awareness, the ability to perceive and act within physical environments. Once intelligence acquires presence, entire industries shift.

Education moves from passive consumption to experiential learning. Healthcare moves from interpretation to real-time assistance. Manufacturing moves from automation to collaboration between humans and intelligent machines. Urban spaces themselves begin to respond dynamically to human behavior.

As with every leap before it, the promise is enormous, and so are the consequences. The difference now is speed and scale. Previous transformations unfolded over generations. Embodied intelligence may reshape systems within years, compressing adaptation timelines and increasing the importance of intentional design.

When Intelligence Shares Our Space

The moment intelligence occupies physical space, trust becomes visceral. A recommendation on a screen can be ignored, but guidance in your field of view feels authoritative. A robotic assistant moving beside you triggers instinctive reactions: comfort, caution, or fear. This raises profound questions:

- Who controls what an embodied AI sees?
- Who decides how it intervenes?
- How do we design consent in shared environments?

Unlike software tools, embodied systems shape perception itself. They influence attention, behavior, and decision-making in ways that are harder to audit and easier to normalize.

Trust must, therefore, be designed into the experience, not added afterward. Transparency, control, and clear boundaries become essential when intelligence shares not just our tasks, but our physical reality.

History teaches us that technologies shape perception. Printing presses, radio, and television reshape culture as much as they reshape productivity. Embodied AI will do the same, but with far greater intimacy.

Learning, Work, and the End of Abstraction

One of the most powerful implications of embodied intelligence is its effect on learning.

Humans learn best through experience, not instruction. Yet modern education has been dominated by abstraction: textbooks, lectures, and exams detached from real-world context.

AI-driven immersive systems reverse this. A student can explore a molecular structure from the inside. A trainee can rehearse emergency scenarios repeatedly without risk. A professional can practice rare, high-stakes decisions in simulated environments that adapt in real time.

The result is not just faster learning, but deeper understanding. Knowledge becomes something we interact with, not just something we memorize. Skills are developed through repetition in context, not through passive exposure.

Work evolves as well. Knowledge workers increasingly collaborate with AI that understands spatial context: designing, diagnosing, building, and repairing in partnership with intelligent systems that see what they see.

The boundary between "digital work" and "physical work" begins to dissolve. What emerges is a new category: augmented work, where intelligence enhances human capability in real time.

Ethical Weight in Physical Form

When intelligence becomes embodied, ethics cannot remain abstract. A biased algorithm is harmful. A biased system acting in physical space is dangerous. Embodied AI systems raise new moral dilemmas:

- How do we prevent manipulation in immersive environments?
- How do we protect privacy when perception itself is automated?
- Who is accountable when a physical system makes a harmful decision?

These questions echo classic philosophical dilemmas but with modern urgency. Responsibility no longer ends at code. It extends into design, deployment, and oversight of systems that act alongside humans.

The closer technology gets to our bodies and environments, the more important it becomes to align it with human values. Ethics must move from policy documents into product design, shaping how systems behave in real-world contexts.

Making Peace with Embodied Intelligence

Fear is a natural response when intelligence steps closer to us. But history suggests that avoidance is not protection. The printing press frightened authorities. Electricity terrified cities. Early automobiles were considered reckless.

What matters is agency. Societies that shape technology deliberately through education, governance, and ethical design benefit from it. Those that react passively are shaped by it.

Embodied AI is not about replacing humans. It is about expanding how intelligence participates in the world. The future will belong to those who learn to coexist with intelligence that shares their space, guided by human judgment, values, and responsibility. This coexistence requires not just acceptance, but intentional integration, designing systems that enhance human dignity, capability, and autonomy.

Closing

Intelligence is no longer content to remain behind glass. It is stepping into our world: into our classrooms, workplaces, hospitals, and cities. Not as a distant tool, but as a presence.

The question before us is not whether embodied intelligence will arrive. It already has. The real question is whether we will shape it with wisdom or allow it to shape us by default. Because in a world where intelligence shares our space, the ultimate outcome will not be determined by the technology itself but by the choices we make in how we design, deploy, and live alongside it.

About the Author

Sriya Burra is a sophomore studying computer science at a top-ranked public Ivy League university, where she merges technical rigor with a commitment to social impact. She previously served as director of technology and programs at The Stemistry Project, a student-led nonprofit delivering STEM education and mental health resources to underrepresented communities. Her contributions to technology and advocacy were recognized with the 2023 NCWIT Aspirations in Computing Rising Star Award from the San Francisco Bay Area Affiliate. Sriya brings hands-on expertise in Python, SOLIDWORKS, and NetLogo, supported by professional certifications in Python and Data Science. As co-creator of an AI-powered educational platform, she has applied her engineering skills to build tools that provide students with personalized exam preparation, AI tutoring, and scenario-based practice, making high-quality business competition prep more accessible to every student.

LinkedIn: https://www.linkedin.com/in/sriya-burra/

Email: sriya12.burra@gmail.com

CHAPTER 17

HOW C-SUITE LEADERS CAN BUILD AI THAT ACTUALLY DELIVERS ROI

By Maribeth Martorana, CFA
Founder & CEO, Sapient Advisors: AI, Innovation
Miami, Florida

Activity is not output.

—Andy Grove, Former CEO of Intel

Most organizations believe they are implementing AI. They are not. They are buying tools, launching pilots, and filling board decks with AI initiatives. Meanwhile the gap between AI activity and AI results grows wider every quarter. This technology is real. The investment is real. But for many leadership teams, the return remains elusive, and the path forward feels anything but clear. I call this the AI Swirl.

Part of what makes the "AI Swirl" so disorienting is that this is not a normal technology adoption cycle. Organizations are not simply adding AI to how they work, they are being asked to reconstruct how they operate. It is what happens when hype and misinformation flood the market faster than leaders can effectively evaluate, when the pace of change creates pressure to act before there is clarity, when executives lack the foundational AI literacy to ask the right questions of vendors or guide their own teams, and when the fear of falling behind drives organizations to chase the next tool before the last one delivered any value. It is not one problem. It is four forces hitting at once, and the cumulative effect is exhausting.

If you are a C-suite executive navigating this right now, you already know what it feels like. The vendor calls all sound the same. The board members are asking about your AI strategy. The team is looking to you for direction while you are still trying to separate signal from noise. The nagging sense that your competitors are moving faster lingers.

As our CTO at Sapient Advisors observes: the technology issues are the easy part. The people issues are the hard part. And most organizations are not yet having the right conversations about either.

The answer is not another tool. It is not another pilot. And it is not strategy alone. Executives have been handed AI strategies that sit in drawers. What delivers results is the right strategy, built on honest organizational readiness, and executed with the discipline to carry it all the way through to measurable outcomes.

Why AI Stalls?

The most common question I hear from executives is some version of this: we have invested in AI. Why isn't it working?

The instinct is to blame technology. Or the vendor. Or the team. In the organizations I advise, failure is almost never where leaders think it is.

AI pilots stall for predictable reasons. Ownership disappears when a pilot moves toward scale. Workflows go untouched, so AI lives alongside the work rather than inside it. Data that looked clean in a demo falls apart under real conditions. Governance arrives too late, and suddenly leadership cannot explain or defend a system they were ready to scale.

Each of these is a structural failure. Not a technology failure. And beneath all of them runs something deeper. AI does not create organizational problems. It exposes them. It also forces a confrontation that many leadership teams have been deferring. The operating layer of the organization, how decisions are made, how workflows, how value is created, is being replaced. That is not a project. It is a transformation.

Broken processes surface. Data gaps become visible. The people and culture questions that were never fully resolved, who owns what, who trusts what, who is willing to work differently, suddenly matter enormously. Employees are not resisting the technology. They are navigating real uncertainty about their roles and their future. That is a leadership problem. It requires a leadership response.

Capital compounds the challenge in ways most executives do not anticipate. Organizations that fund AI as a project, with a start date, an end date, and a budget that expires, will not scale it. Once AI moves into the operating layer of the business, it must be funded and governed like any core operational capability. One-time thinking produces one-time results, or none. Strategy alone will not save you.

Building the Right Strategy

The right strategy delivers results combined with the discipline to execute it, through real workflows, real data, real people, and real accountability, all the way to a measurable outcome. That strategy begins not with technology but with a question. What business problem are we trying to solve? It sounds obvious. But in most organizations, this question never gets answered clearly before AI tools are purchased and pilots are launched. The result is investment without direction. Activity without output.

In working with clients, one should work backwards from four possible outcomes: increasing revenue, improving operational efficiency, reducing risk, or improving customer experience. Before any use case is defined, before any vendor is evaluated, leadership must be able to name which of these they are pursuing and how they will measure progress. If that answer is not clear, the organization is not ready to invest in AI. It is ready to invest in clarity.

Once the business objective is defined, use cases can be prioritized deliberately. Not every idea is worth pursuing. The discipline is in choosing the use cases that balance feasibility, business value, and organizational readiness, and then validating them through structured design sprints before committing to a full build. Pilots should generate learning, not just momentum. Which brings us to readiness. This is where most AI initiatives tell the truth about themselves.

Readiness is not a checklist. It is a leadership conversation across five dimensions.

The first is data. Is your data clean, owned, and defensible enough to support AI-informed decisions? If the answer is no, data readiness is the first project, not an afterthought.

The second is infrastructure. Can your systems support what you are building in terms of integration, compute, and security? This is the foundation everything else runs on.

The third is people and talent. Do your teams have the skills to work alongside AI? Have you defined how roles will change, not just for technical staff but for every person whose work AI will touch? This is not a training question. It is a workforce design question, and it needs to be answered before deployment, not after confusion sets in.

The fourth is culture. Is there genuine psychological safety to experiment, fail, and iterate? Do employees trust that AI is being introduced to augment their work rather than eliminate them? Leaders set this tone. It determines whether adoption accelerates or quietly stalls. You cannot mandate a culture of AI readiness. You have to earn it.

The fifth is capital allocation. This is not a budget question. Once AI moves into the operating layer of the business, investment must be sustained. The true cost includes compute, data preparation, governance, and the ongoing human judgment required to monitor and adapt. Organizations that treat AI investment as a one-time project will not scale it. Those that treat it as an operational commitment will.

Executing with Discipline

Strategy tells you where to go. Execution is how you get there. But here is what most firms miss. Executing an AI strategy is not the same as implementing a technology. It is reconstructing how an organization operates.

The operating layer of the business is being replaced. Workflows are being redesigned. Decision rights are being redrawn. Business models that were built around human-only processes are being rebuilt around human and intelligent systems working together. This is not an upgrade. It is a structural transformation. And the leaders who understand that will make fundamentally different decisions than those who do not.

That is why execution requires a different kind of commitment than strategy work. Most firms approach it as implementation; they deliver the roadmap, hand it to a project team, and move on. What gets left behind is an organization that knows what to do but has not been built to do it. The strategy is sound. The operating model is not yet ready to carry it. Execution means staying in the room. It means five disciplines, applied consistently, until measurable results are in hand.

The first is embedding AI into real workflows, not alongside them. AI that lives outside the user workflow will be ignored, even when the outputs are accurate. Redesigning workflows is not a technological task. It is an organizational design task, and it requires leadership commitment, not just technical capability.

The second is governance, built in from day one. As decision rights shift between human and intelligent systems, the boundaries must be explicit. Where does AI recommend and a human decides? Where can AI act? Where must a human always be in the loop? These are not compliance questions. They are operating model questions, and they must be answered before deployment, not after a problem forces the conversation.

The third is designing human judgment at every step. As roles change and intelligent systems take on more of the operating layer, people need clarity about where their judgment begins and ends. When that clarity is absent, trust erodes and user adoption stalls. You cannot reconstruct an operating model without bringing the people inside it with you.

The fourth is measuring outcomes, not activity. The question is never how many pilots are running. It is which business metric has moved. If you cannot point to a number, you have not executed. You have experimented.

The fifth is treating AI as a living system, not a completed project. The operating layer you are building today will need to adapt as data changes, user behavior evolves, and technology advances. Organizations that build continuous monitoring and governance into the model from the start will compound results over time. Those that treat deployment as the finish line will not. Execution is where reconstruction either happens or stalls. There is no middle ground.

What This Looks Like in Practice

The principles in this chapter are not theoretical. JPMorgan Chase has lived them at scale. When generative AI emerged, most financial institutions defaulted to a wait-and-see posture. JPMorgan did not. But what distinguished the bank was not simply that it moved early. It was how it moved.

Leadership started with clear business objectives, not a tool evaluation. A strategic mandate focused on the domains where transformation would have the greatest impact: credit, fraud, marketing, operations, and frontline banker enablement. The question was never which AI to adopt. It was where AI would most change the trajectory of the business.

Data readiness was treated as non-negotiable. People and culture were designed into the strategy from the beginning, not bolted on afterward. JPMorgan took a learn-by-doing approach, putting tools directly in the hands of employees. New analyst hires received mandatory training in prompt engineering. Entirely new job categories were created, including context engineers, knowledge managers, which did not exist before. The operating model was being reconstructed, and the workforce was being rebuilt alongside it. Capital was committed as an operational investment, not a project budget. AI benefits grew thirty to forty percent year over year.

The results followed. The Coach AI system increased adviser productivity by ninety-five percent and contributed to a twenty percent increase in sales between 2023 and 2024. By 2025, the bank was generating two billion dollars in annual business value from AI, up from one hundred million just three years earlier. Technology was not the hard part. Building the strategy, the readiness, and the discipline to execute it all the way through was.

Your Results Guide

The AI Swirl is real. But it is not inevitable. Here are six things you can do right now to move from activity to results.

1. **Stop shopping.** Before evaluating another vendor, write down the one business outcome you are trying to change and the metric you will use to measure success. If you cannot answer that question clearly, you are not ready to buy anything. You are ready to get clear.
2. **Run a data reality check.** Ask your team honestly: if we deploy AI tomorrow on our highest priority use case, could we defend the data behind every output? The answer will tell you where to invest first.

3. **Assess your people and culture readiness before you deploy.** Define how roles will change. Communicate clearly and early. Create the conditions for employees to engage with AI without fear of blame or replacement. The technology issues are the easy part. This is the hard part. Design it accordingly.
4. **Bring the right people into the room from day one.** AI decisions touch every function. If only the CTO is part of the conversation, you are designing for resistance. Operations, compliance, risk, and the people who do the work all need a seat at the table before decisions are made.
5. **Build governance before you need it.** Define decision rights, escalation paths, and override protocols before your pilot scales. Governance is not a brake on AI. It is what gives your organization the confidence to move fast without losing control.
6. **Restructure how you fund AI.** If AI is still a project line with a start date and an end date, you have already constrained its potential. Decide whether AI is part of how your organization operates. If so, then fund it like it is.

Closing

AI is not coming. It is already here, already reshaping how organizations operate, already separating the leaders who are building something durable from those who are still running experiments.

The organizations that win will not be the ones that bought the most tools or launched the most pilots. They will be the ones that had the discipline to ask the right questions first, honesty to assess their true readiness, and the commitment to execute all the way through to results. The operating layer of your organization is being reconstructed whether you lead that process deliberately or not. The only question is whether you are designing it or reacting to it. Most firms will help you build the plan. Far fewer will stay through the execution. That distinction is everything.

About the Author

Maribeth Martorana is the Founder and CEO of Sapient Advisors, a data and AI consultancy helping financial services, insurance, and commercial real estate companies turn AI ambition into measurable business results. With over 25 years of experience across finance, data, and product strategy, she specializes in guiding C-suite leaders from AI experimentation to execution and ROI.

Maribeth partners with organizations to assess AI readiness, prioritize high-impact use cases, and deliver de-risked prototypes and MVP roadmaps through structured design sprints. Her expertise spans the full data lifecycle, from sourcing and structuring machine-readable data to aligning data products with monetization and scalable AI deployment.

Previously, Maribeth held senior leadership roles at Aiera, Moody's Analytics, CFA Institute, and FactSet, where she led AI-driven product innovation, commercialization, and go-to-market strategy. She focuses on helping organizations cut through AI noise and translate strategy into measurable business impact.

Email: Maribeth@sapient-advisors.com

Company website: www.sapient-advisors.com

Individual website: https://maribethmartorana.substack.com

LinkedIn: https://www.linkedin.com/in/maribeth-martorana/

CHAPTER 18

CLARITY IS THE REAL AI OPPORTUNITY

By Leonardo Veri
Founder, CTO, and AI Advisor
São Paulo, Brazil

We are living through civilization's next great disruption. Artificial intelligence is no longer a laboratory experiment or a Silicon Valley fantasy. It is infrastructure. It is embedded in software, logistics, healthcare, marketing, defense, education. It is rewriting how value is created, captured, and distributed. And yet, something strange is happening.

As intelligence becomes more accessible, organizations seem more confused than ever. Dashboards multiply. Models become more sophisticated. Agents automate workflows. Yet strategic paralysis increases. Leaders hesitate longer. Teams debate endlessly. Investments grow, but conviction shrinks.

The AI opportunity is real. But so is the AI illusion. The illusion is believing that intelligence alone creates advantage. It doesn't. Clarity does.

The Core Mistakes: Starting with Solutions

Every solution needs a problem. It sounds obvious. Yet in the AI era, this principle is violated daily. Executives announce "AI initiatives" before articulating what they are truly trying to solve. Startups integrate large language models into products without understanding the decision architecture of their customers. Boards demand automation without defining the trade-offs they are willing to accept.

Technology becomes theater. Innovation becomes optics. AI becomes a badge.

As Naval Ravikant writes in *The Almanack of Naval Ravikant*, "The hardest thing is not doing what you want. It's knowing what you truly want."[1] The same is true for organizations. The hardest thing is not building the solution. It is defining the problem.

In 2016, long before generative AI became mainstream, I developed a framework called Frame The Challenge. It emerged from observing how teams repeatedly jumped to solutions without mapping context. We would slow down. Map stakeholders. Surface incentives. Identify constraints. Model second-order effects. Ask uncomfortable questions. Only after exhausting the context would we attempt to define the problem, ideally in one simple and clear sentence. That sentence was always harder to write than the entire solution roadmap. And it changed everything.

Industry research, including findings referenced in the State of AI in Business 2025 Report, indicates that a significant majority of AI initiatives fail to deliver measurable business value, often because they are not anchored in clearly defined problems.[2]

AI does not fail at execution. It fails at definition. When the problem is vague, AI amplifies vagueness. When the problem is misaligned, AI scales misalignment.

The technology is not the bottleneck. Clarity is.

From Information Scarcity to Meaning Scarcity

For most of human history, power belonged to those who controlled information. Kings controlled messengers. Corporations controlled data. Experts controlled access to knowledge.

Today, intelligence is abundant. You can generate strategic reports in seconds. Simulate financial forecasts instantly. Produce code, marketing campaigns, legal drafts, and creative assets at scale. We solved information scarcity. Now we face meaning scarcity.

When everyone has access to intelligence, advantage shifts. It no longer belongs to those who compute faster. It belongs to those who orient better. When intelligence becomes abundant, judgment becomes premium.

In a world where machines can answer almost any question, the quality of the question becomes the true differentiator. Most organizations are drowning in outputs but starving for direction. They have more data than ever and less shared understanding. They have predictive models and no alignment. They have automation and no clarity about where they are heading.

The shift of AI is not acceleration. It is orientation.

The VeriLabs Compass: Navigating the Disruption

If AI is wind, clarity is direction. Wind without direction accelerates drift.

At VeriLabs, we operate as an AI Venture Studio. That does not mean we simply integrate tools or build prototypes. It means we design ventures, products, and decision systems with AI at the core but anchored in orientation. Over time, this approach evolved into what we call the VeriLabs Compass. It is not a technology framework. It is a clarity framework.

1. True North: Problem Clarity

Before discussing tools, we ask a harder question: What are we truly solving and which single decision, if dramatically improved, would change our future?

Most teams struggle here. They describe symptoms instead of problems. They describe features instead of decisions. They describe goals instead of trade-offs.

True North is uncomfortable because it forces prioritization. It forces exclusion. It forces commitment. Without it, AI initiatives float.

2. Reality Mapping: Context and Systems

Problems do not exist in isolation. They live inside systems. Markets. Incentives. Regulations. Customer behavior. Internal politics. Cultural biases.

AI systems trained without systemic awareness optimize locally while damaging globally. Reality mapping expands the lens and asks: What system are we truly operating within and who wins, who resists, and what invisible constraints or risks shape the outcome?

Clarity requires context.

3. Decision Architecture

Organizations are decision factories. Strategy is a set of decisions under uncertainty. Execution is a chain of micro-decisions. Yet few companies explicitly map their decision layer. So we ask: Which decisions actually drive disproportionate value and are they clearly owned, informed, and governed by explicit trade-offs?

Instead, many implement AI to automate tasks without redesigning the decision layer. This is backwards. AI should strengthen decision architecture, not bypass it

4. AI at the Core

Only after orientation is clear does AI enter. Here, AI becomes powerful:

- Reducing noise
- Surfacing patterns
- Simulating futures
- Modeling scenarios
- Stress-testing assumptions

AI is not the compass. It is the wind. With direction, wind creates speed. Without direction, it creates chaos.

5. Execution and Feedback

Clarity is not static. It is iterative.

AI systems generate signals. Those signals must inform decisions. Decisions must generate outcomes. Outcomes must refine the model. Feedback loops create learning. Learning refines clarity. This is not about digital transformation. It is about cognitive transformation.

From AI Adoption to AI Orientation

Most companies are focused on adoption: What tools should we use? What model should we integrate? How do we automate this workflow? These are tactical questions.

The strategic question is different: Which decision, if dramatically improved, would redefine our trajectory?

Adoption is about integration. Orientation is about transformation. AI adoption without orientation produces marginal productivity gains. AI orientation redesigns how value is created.

In an AI Venture Studio model, we ask a more radical question: If this company were born today, AI-native from day one, what would look fundamentally different? What would be automated by default? What decisions would be modeled instead of debated? What value proposition would be reimagined? What second-order consequences would we design for instead of react to?

The future does not belong to those who add AI to legacy systems. It belongs to those who redesign systems around AI and clarity, simultaneously.

The Rise of the Clarity Architect

Corporate roles evolve with technological waves.

The CIO managed infrastructure.

The CTO managed technology.

The CDO managed data.

The CAIO manages AI.

But something deeper is needed now. Organizations need someone who owns the problem definition. Someone who disciplines thinking before scaling intelligence. Someone who ensures semantic alignment across teams. Someone who can translate between business, technology, and human incentives. Call it "chief clarity officer." Call it "clarity architect."

The title is irrelevant. The function is essential. In the AI era, leadership is less about control and more about cognitive discipline. It is about asking: Are we solving the right problem, and are we aligned on the trade-offs and reality that define it?

Machines compute. Leaders orient.

What Individuals Must Develop to Survive the AI Era

The AI disruption is not only corporate. It is personal. Many ask: What skills should I learn? Should I learn to code? Should I become a prompt engineer?

Coding is valuable. Technical literacy matters. But something more durable is emerging. The safest career in the AI age is not coding. It is learning how to define meaningful problems and asking the right questions. Three capabilities will remain defensible:

1. Problem Framing Literacy

The ability to turn ambiguity into clarity. To reduce noise into a single, precise problem statement. To distinguish symptoms from root causes. To make trade-offs explicit instead of implicit.

Those who frame problems shape solutions. Those who shape solutions shape outcomes. In a world flooded with answers, the rare skill is asking the question that matters.

2. Systems Thinking

AI does not operate in isolation. Neither do markets. Neither do careers. Every decision lives inside a system of incentives, feedback loops, constraints, and unintended consequences. Linear thinkers look for direct cause and effect. Systems thinkers look for second-order impact. They ask: If this works, what changes next? If we optimize here, what breaks elsewhere? Who adapts? Who resists? What shifts quietly over time?

In an interconnected world amplified by AI, local optimization can create global fragility. Those who understand systems do not chase outputs. They design for resilience.

When machines accelerate complexity, the human advantage is seeing the whole board.

3. Decision Confidence Under Certainty

AI produces probabilities, not certainty. Models generate forecasts. Simulations project scenarios. Dashboards quantify risk. But no model eliminates uncertainty.

At some point, a human must decide. The defining skill of the AI era is not prediction. It is commitment. Can you act without perfect information? Can you integrate data with judgment? Can you accept trade-offs without paralysis?

The future will not reward those who wait for clarity to be complete. It will reward those who build enough clarity to move.

Machines reduce ambiguity. Leaders absorb it. In a world of intelligent systems, courage becomes strategic.

The New Competitive Equation

The AI shift is real. It will reshape industries, compress timelines, democratize intelligence, and accelerate innovation at a scale we have never seen before. But intelligence alone does not create advantage.

Many organizations will deploy sophisticated AI systems and remain strategically lost. Others will deploy modest AI and outperform because they know exactly what they are solving. The difference will not be technical maturity. It will be cognitive maturity.

The new competitive equation is simple: Problem Clarity + Decision Architecture + AI Amplification = Sustainable Advantage

Or even more simply: Clarity × Speed × Courage = AI Advantage

When intelligence becomes abundant, clarity becomes scarce. When answers are infinite, the right question becomes priceless. In civilization's next great disruption, the winners will not be those who build the smartest machines. They will be those who define the right problems, design better decisions, and commit with conviction.

The wind is rising. You can build bigger sails. Or you can choose your direction. In the AI era, that choice, that clarity is the real advantage.

References

1. Eric Jorgenson, *The Almanack of Naval Ravikant: A Guide to Wealth and Happiness* (Magrathea Publishing, 2020).
2. MLQ.ai, *State of AI in Business 2025 Report* (MLQ.ai, 2025), https://mlq.ai/media/quarterly_decks/v0.1_State_of_AI_in_Business_2025_Report.pdf.

About the Author

Leonardo Veri is the founder of VeriLabs, an AI venture studio dedicated to helping organizations navigate complexity and build AI-native businesses with clarity at the core. He works with CEOs, founders, and executive teams to transform technological disruption into structured decision advantage.

With over two decades of experience across technology, innovation, and enterprise strategy, Leonardo has advised organizations ranging from large enterprises and traditional SMBs to high-growth startups. His work focuses on problem framing, decision architecture, and the strategic integration of AI as a lever for sustainable value creation.

Through his framework-driven approach, including the VeriLabs Compass and Frame The Challenge methodology, he helps leaders move from confusion to conviction in the age of artificial intelligence.

Leonardo is also a keynote speaker, global mentor, and host of the *VeriTalks* podcast, where he explores leadership, AI, and the future of decision-making.

Email: leonardo.veri@gmail.com
Website: www.leoveri.com
LinkedIn: https://www.linkedin.com/in/leonardoveri

CHAPTER 19

THE HUMAN LINCHPIN: THE BETTER AI GETS, THE MORE VALUABLE YOU BECOME

By Rafa Maximo
AI Marketing; Creator of The Human Linchpin System
Miami, Florida

We know more than we can tell.
—Michael Polanyi

May 18th, 2025. It should be another regular Sunday evening in Stockholm. This one, though, is a bit different. For Klarna's shareholders, it is the day before the company announces its first quarter results, much anticipated given its heavy investment and push in AI innovation. For Klarna's CEO, Sebastian Siemiatkowski, this earnings call has a different ring to it, not only because the company has marked its fourth consecutive profitable quarter, reaching 100 million active users while reducing its workforce by roughly 40% since 2022, with AI central to that transformation. To Sebastian, this time is different because he is about to do something extraordinary: be in two different places at the same time.

With the wonders of artificial intelligence, Klarna is outsourcing its own CEO's presence, delivering the company's results through an AI-generated avatar built in Sebastian's likeness. The avatar presents the numbers. The real Sebastian is out there building rapport with key investors, observing his team, creating the kind of trust that only physical presence can build. One might think it theatrical. Quite the opposite. They are doing exactly what they are selling: the advantages of AI at scale.

But here is the question nobody in that earnings call thought to ask: When the person responsible for building trust in a company outsources his own presence, what, exactly, is the company selling?

Klarna has also been one of the companies most openly sharing how AI has reduced their human workforce, reigniting a familiar debate: Will artificial intelligence replace human professionals? That question is understandable. It creates urgency, headlines, and existential dread. But it is also a distraction. It keeps the conversation about the tools, not the brands. And that distraction is expensive.

MIT's State of AI in Business 2025 reported that only about 5% of task-specific enterprise GenAI tools were successfully implemented, with the authors describing a 95% failure rate for enterprise AI solutions. Not because the technology doesn't work. Because the central problem lies in the absence of strategy, organizational maturity, and leadership capable of transforming technology into real business value.

A Multiplier Has No Judgment

A multiplier amplifies what is already there. Put strategic clarity in, and you get more strategic clarity out. Put a hollow brand promise in, and you get that promise at scale, with better production value, at lower cost, indistinguishably. So the question that actually matters is harder and less comfortable: What happens to a brand when it starts optimizing for efficiency at the expense of the things that made people trust it in the first place?

Artificial intelligence can generate a strategy. That needs to be said directly. Fed the right inputs, original thinking, genuine cultural insight, and real brand identity, it produces extraordinary output. This is not a warning about a weak tool. The tool is genuinely powerful.

But here is what that power conceals: AI generates from inputs. The quality of what it produces is entirely dependent on the quality of what humans feed it. Which means the moment brands stop doing the hard human work of developing real insight, the output homogenizes. Differentiation disappears. Quietly. Gradually. Comfortably.

Disney Pixar made this argument in 2008 and dressed it as a children's movie, so nobody took it seriously as a business warning. In *WALL-E*, the humans didn't lose the ability to walk because someone took walking from them. They lost it because the system was so good at moving them that they never needed to practice anymore. The atrophy was invisible. Gradual. Comfortable. And by the time it mattered, when the ship needed humans to stand up and take control, they had forgotten how. The humans in *WALL-E* didn't choose atrophy. They chose comfort. Those are different things. And only one of them feels like a warning while it is happening.

This is us. Just earlier in the process. We are not choosing to remove our souls from brands and products. We are choosing the convenience of doing things faster, more easily, more efficiently. The problem emerges when we expect AI to be in the driver's seat, piloting the ship. When AI is our navigator, our onboard computer, it is extraordinary, guiding us toward informed decisions, surfacing insights, processing data at speeds no human team can match. The moment we change seats, we begin losing something we don't notice losing until it is gone. We won't just forget how to fly. We will forget why the destination mattered.

The strategic consequence plays out at the market level too. We expect AI to deliver personalization: the right person, right message, right moment. But if every brand optimizes through the same logic, feeds the same tools, chases the same signals, the algorithm doesn't produce a winner. It produces a tie. You reach the right person, at the right moment, with the same message as everyone else. That is not personalization. That is targeted sameness. And in competitive markets, sameness doesn't convert. The algorithm doesn't know what makes you different. Only you do. Stop telling it, and it will make you the same as everyone else, efficiently, at scale, while the dashboard shows green.

Audiences have been trained. Fifteen years of influencer culture, bot accounts, deepfakes, and algorithmic content have made people, especially younger generations, acutely sensitive to the absence of genuine human decision-making behind content. They cannot always name what feels wrong. But they feel it.

Coca-Cola illustrates what happens when brands miss this signal. In late 2024, they reimagined the iconic "Holidays Are Coming" campaign—same concept, same music—fully produced by generative AI. The backlash was immediate. Consumers described the result as unnatural, uncanny, off. But the deeper problem was not the technology. It was the contradiction.

Coca-Cola built its entire relevance around a slogan that reads, 'It's the real thing.' Then it produced one of its most iconic campaign moments in a format that was, by definition, entirely artificial. The audience that had grown up watching the famous red truck roll into their cities, in real life, building the warmth and joy the brand embodies, watched all of that become simulated. The craft was absent. The intent behind the craft was absent. The result was Brand Authenticity Collapse: AI execution directly contradicts the brand promise. The technology succeeded. The brand failed. A multiplier had done its job perfectly. It multiplied the wrong thing. Coca-Cola faced the backlash and chose the same route the following year. Whether that reflects strategic conviction, experimentation at scale, or a misreading of what audiences actually value is still unclear. What is clear is the tension it exposed: When a brand built on "the real thing" embraces artificial production, the audience doesn't debate the technology. They question the brand. Once that happens, you're no longer optimizing performance. You're repairing trust.

Klarna tells a different dimension of the same story. The company replaced hundreds of customer service employees, celebrated the efficiency gains publicly, then quietly began rehiring as customer experience deteriorated. They handed the wheel to the autopilot and discovered that some moments require a human hand. In response, they launched a CEO hotline, Sebastian's avatar, available to answer client questions as an AI bot. The real Sebastian, meanwhile, is out doing what the avatar cannot: building trust through presence.

Aerie, the apparel brand owned by American Eagle Outfitters, made a different bet and made it early. In 2014, the brand pledged to stop retouching images of its models. It worked. A decade later, facing the same generative AI pressure every other brand was navigating, Aerie extended the same logic: no AI-generated people or bodies in any campaign imagery. Its message was clear: "100% Aerie real."

The results were unambiguous. Their pledge post became the brand's most liked Instagram post. In Q4 2025, sales rose more than 20%. Stacey McCormick, Aerie's CMO, explained the reasoning without jargon: "In an industry where everything is generated, realness becomes special."

That sentence is the sharpest strategic insight in the current AI marketing conversation. When a capability becomes ubiquitous, its opposite becomes scarce. AI made simulation cheap and abundant. That made authenticity expensive and valuable in a way it has not been since before digital retouching made perfection the default. Aerie didn't win by rejecting technology. They won by knowing exactly what not to automate, by understanding what their brand was actually built on, and protecting that asset before the pressure to automate it arrived.

This dynamic is not uniform across markets, and that is a dimension most AI consultants miss entirely. Having built and led learning programs across Brazil, the United States, and Europe, I have seen firsthand how differently audiences in each market respond to automation in brand communication. What reads as modern efficiency in one context reads as cold indifference in another. A global

AI strategy that treats this variance as irrelevant is not a strategy. It is an assumption. And assumptions, at scale, are expensive.

Marketing has always been the discipline of translating human experience into meaning that moves people. Not data into content. Experience into meaning. That distinction determines what kind of intelligence the job actually requires.

Human Knowing Operates in Layers

At the base sits something older than language, inherited patterns nobody taught and nobody remembers learning. Carl Jung called these "archetypes": the hero, the caregiver, and the outlaw. Brands use them constantly, often without naming them. They work because they tap into something audiences recognize before they are consciously aware of recognizing it. Nobody knows where those recognitions began. They are simply known because they are part of what we are.

On top of that sits what philosopher Michael Polanyi called "tacit knowledge": "We know more than we can tell." This is the knowledge that moves through presence, through watching, through being in the room when something important happens. How to read when trust is breaking before anyone says a word. When a campaign feels dishonest before you can prove it. When a room has shifted. Nobody documented this. It was passed through embodiment, not documentation.

And on top of that, the real-time layer: embodied sensing. The instinct that fires before the data arrives. In research on virtual reality, a participant described the experience with precision that stays with you: "My mind knows it's not real, but my body doesn't." That gap between cognitive knowing and embodied feeling is where human experience actually lives. A VR simulation of a complicated birth made students laugh and made one person in the room feel visceral anxiety, because that person had actually been in a delivery room and had felt what that moment cost. The students weren't callous. They simply had no embodied reference.

AI is permanently in that position. It has read every birth story ever written, but it has never been in the room. AI reads signals. Humans read meaning, assembled from all three layers simultaneously. That is not a technical limitation waiting to be solved with more compute or better models. It is the permanent condition of a system that has never been embodied, never inherited anything, never been in the room.

AI Doesn't Have Guts

It can calculate risk. It cannot take responsibility for it. It cannot decide to take it anyway because something human—something older than language, passed through generations of presence—told you the data was missing something important. AI is not afraid of losing a client. It does not feel the weight of a decision. Its livelihood does not depend on getting it right. This is where most brands quietly start destroying their differentiation.

Seth Godin famously argued that people need to become linchpins: indispensable to their organizations. In the age of AI, the question shifts. It's not just which people are irreplaceable, but which parts of what a brand does are too human to hand over. Not everything in marketing should be treated equally in the AI era. Some tasks should be automated: the repetitive, the process-driven, the operationally predictable. Some should be co-piloted, where AI expands thinking and human judgment steers. And some, the parts that shape trust, meaning, symbolic coherence, and genuine differentiation, should never be fully outsourced. I call this last zone the Human Linchpin Factor. Not the anti-AI position. The pro-brand position. And it is where the real strategic work begins.

Here is the risk that the current AI and marketing conversation is not naming precisely enough, and it deserves to be stated as the thesis it is. AI researchers call a version of this "model collapse," when models trained on AI-generated data progressively lose the diversity and authenticity of the original human signal. What I'm describing is its brand equivalent: the same degradation happening not to a model's outputs, but to a company's relationship with its actual customers.

AI trains on human behavior. Brands use it to predict, target, and personalize. Humans consciously and unconsciously begin performing for algorithms, creating content optimized for engagement signals, behaving in ways the model rewards. AI trains on that behavior. The authentic human signal in the data gets progressively diluted until you're optimizing against a version of your customer that no longer exists. Future models train on a version of humanity that is increasingly a reflection of what previous models produced. Brands built on genuine human connection start reflecting a synthetic version of their customer, back at them. At the end of that loop: brands predicting constructed behavior, targeting manufactured preferences, speaking to an algorithmically assembled ghost of their actual customer.

Nobody sounds the alarm. The metrics kept improving the whole way down.

Most consumers will not audit this consciously. But they will feel the brand growing hollow, in ways they cannot articulate but act on, at the moment of purchase, or the quiet moment they choose someone else instead.

Think of organic produce. Craft beer. Artisan bread. Vinyl records. Every industrialization creates the same dynamic: the mass market goes automated and stays there, and in every case, the human version becomes rare, not dominant, but premium. Not for everyone, but for the customers who still know the difference. AI is the newest industrial revolution. Human-made, human-thought, human-felt, human-decided, is becoming the newest artisan product.

Sebastian Siemiatkowski went all-in on AI, replacing even himself with an avatar. But he is not sitting in a command center, removed from the world. He is out there making connections, talking to people, feeling the texture of the business he runs. He outsourced his image, his speech. Not his presence.

Stacey McCormick built Aerie's creative strategy on realness. She is almost certainly using AI to optimize systems, communications, and processes. She outsourced the repetitive. Not the irreplaceable.

Both understood something most brands are still working out: The question is never whether to use AI. It is knowing what you are protecting while you do.

The brands that will lead the next decade are not the ones that adopted AI first. They are the ones that kept their hands on the wheel long enough to know what was below the clouds and had the guts to protect it.

AI will make content infinite. What it cannot make is meaning. That is still a human job. And in a world of infinite content, it is the only one that matters.

About the Author

Rafa Maximo is an AI Marketing and Digital Growth Advisor, creator of a strategic framework for applying artificial intelligence without eroding what makes brands valuable. He held roles at IBM and WPP before spending six years at Meta, where he helped build programs that reached over 20 million learners and trained more than 23,000 professionals across 16 countries.

He holds a master's in media anthropology, a lens that shapes his work at the intersection of technology and human behavior. He also lectures on AI-driven marketing at graduate programs in Berlin and serves as a judge for the Webby Awards.

Today, he advises companies on AI implementation, brand strategy, and growth, helping them decide not just how to use AI, but what must remain human. He has spoken at eMerge Americas and Adweek, among other international stages.

Email: rafaelpmaximo@gmail.com

LinkedIn: linkedin.com/in/rafaelpmaximo

CHAPTER 20

EMBODIED INTELLIGENCE: THE RISE OF PHYSICAL AI

By Vinodhini Ravikumar
AI Leader and Co-Founder at Vault Voyage Ltd
Seattle, Washington

The machine does not isolate man from the great problems of nature but plunges him more deeply into them.

—Antoine de Saint-Exupéry

Most people today experience artificial intelligence through a screen. You type a question, and it answers. You ask it to write something, and it responds instantly. It feels intelligent, sometimes even human. But behind the scenes, what is happening is much simpler than it appears. These systems break language into small pieces called "tokens" and predict what is most likely to come next based on patterns they have learned. This ability has transformed how we work, communicate, and create. Yet this is only the beginning. The next shift is far more profound. AI is leaving the screen.

Growing Up Building Machines That Could Fail

I grew up at a time when access to advanced technology was limited. We did not have modern labs or expensive equipment. What we had was curiosity. As a student, I worked on small robotics projects using basic components like wires, motors, and sensors. Most of the time, things did not work. Circuits failed. Sensors behaved unpredictably. Systems broke in ways we could not anticipate. But when something worked, it revealed something powerful. These were not just programs running in isolation. These were systems interacting with the real world. That experience taught me something fundamental very early on: The real world is messy. Unlike software systems, which operate in controlled environments, physical systems must deal with uncertainty. Conditions change. Inputs are noisy. Outcomes are not guaranteed. Designing for this world requires a completely different mindset.

From Predicting Words to Taking Action

For years, AI has been largely focused on prediction. It analyzed data, generated text, and produced outputs based on learned patterns. Today, we are entering a new phase. AI is beginning to act. It can see through cameras, interpret signals from sensors, and make decisions in real time. This shift is what I call "embodied intelligence." It is when AI moves from being purely digital to becoming part of systems that exist in the physical world. Physical AI combines software intelligence with sensors, robotics, and real-world infrastructure. It enables machines to perceive, understand, and act. This is not just an incremental improvement. It is a fundamental shift in how intelligence is applied.

Where Physical AI Is Already Transforming Industries

This transition is already underway across industries. In manufacturing, AI systems detect defects in real time with precision beyond human capability. In logistics, intelligent systems optimize supply chains across complex global networks. In transportation, autonomous systems are redefining mobility. In agriculture, AI helps farmers make better decisions about water, soil, and crop health. In energy, AI optimizes grids and reduces waste. These are not future possibilities. They are present realities. What makes this shift so powerful is that the real economy runs in the physical world. Factories, transportation systems, infrastructure, and cities all operate in physical environments. When intelligence becomes embedded into these systems, it changes how they function at a fundamental level.

Why the Real World Changes Everything

Traditional software systems operate in relatively predictable environments. Inputs are structured. Outputs are expected. Failures are manageable. The physical world does not behave this way. Lighting changes. Sensors fail. Environments shift. Humans behave unpredictably. A system that performs perfectly in a lab can struggle in real-world conditions. This is why physical AI is fundamentally different from digital AI. It is not just about building smarter models. It is about building systems that can operate reliably in uncertain environments. This requires a shift from model thinking to systems thinking.

The Invisible Layer That Determines Success

There is a layer of AI that most people do not see but that determines success more than anything else. That layer is infrastructure. Behind every intelligent system is a network of components working together. Data pipelines move information. Compute systems process it. Edge devices bring intelligence closer to where decisions need to be made. Networks ensure connectivity. Security systems protect integrity. In physical AI, this infrastructure becomes critical. Decisions often need to be made in real time. A delay is not just inconvenient. It can be costly or even dangerous. A failure is not just a bug. It can impact real-world outcomes. This is why the future of AI will not be won by models alone. It will be won by those who understand how to design, build, and operate systems.

How Physical AI Systems Work

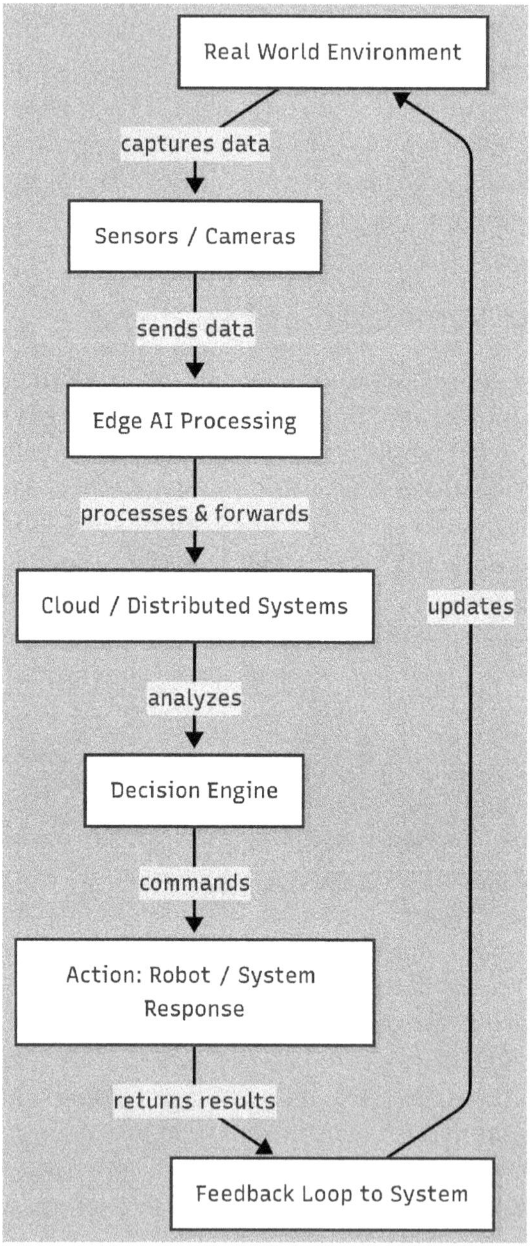

This loop is continuous. Physical AI systems learn, adapt, and improve over time by constantly interacting with the environment.

Physical AI in Healthcare: Where It Becomes Personal

Healthcare has always been an environment defined by urgency, complexity, and consequence. Unlike many other industries, decisions are not abstract or reversible. They directly impact human lives. For decades, technology in healthcare has focused on digitization, making records easier to access, improving communication between systems, and enabling better storage of information. While these advances have been important, they have largely remained within the digital layer. Data is collected, stored, and analyzed, but the actual decision-making process still depends heavily on human interpretation and delayed feedback loops. This is where physical AI begins to fundamentally change the equation.

Physical AI shifts intelligence from being something that is consulted after the fact to something that operates in real time, directly within the environment where care is delivered. It brings together sensors, imaging systems, and on-device intelligence to create continuous awareness. Instead of waiting for a scan to be reviewed hours later, or for a specialist to interpret results in isolation, physical AI enables systems to process information as it is generated. In pathology, for example, this means analyzing tissue samples the moment they are captured, identifying patterns and highlighting anomalies immediately. The goal is not to replace clinicians, but to augment their perception, reducing the cognitive burden and increasing consistency in interpretation.

What makes this shift powerful is not just speed, but reliability. Human decision-making, especially in high-pressure environments, is influenced by fatigue, workload, and variability in experience. Physical AI introduces a layer of consistency that operates regardless of these factors. It does not get tired, it does not lose focus, and it does not overlook subtle patterns due to time constraints. Instead, it continuously evaluates incoming data, providing signals that help guide attention where it matters most. This transforms healthcare from a reactive system into a proactive one, where issues can be identified earlier and acted upon more effectively.

Another critical aspect of physical AI in healthcare is proximity. Traditional AI systems often rely on sending data to external systems for processing, which introduces delays and dependencies. In healthcare, even small delays can matter. By embedding intelligence directly into devices and systems, physical AI reduces this gap. Decisions can be made locally, without waiting for external processing. This not only improves responsiveness but also reduces reliance on continuous connectivity. In environments where infrastructure may be limited or where latency is critical, this becomes a significant advantage.

This localized intelligence also has implications for cost and scalability. Healthcare systems already operate under significant financial pressure. Constantly relying on external processing adds ongoing operational costs that scale with usage. Physical AI allows more computation to happen within the system itself, reducing the need for repeated external interactions. Over time, this creates a more sustainable model, where intelligence is not just powerful but also economically viable. It enables healthcare providers to deploy advanced capabilities without proportionally increasing costs, making innovation more accessible across different settings, from large hospitals to smaller clinics.

Ultimately, physical AI in healthcare is not about replacing human expertise. It is about redefining how that expertise is applied. It allows clinicians to focus on complex decision-making, empathy, and patient interaction, while systems handle continuous monitoring and pattern recognition. It creates a partnership between human judgment and machine consistency, where each complements the other. This is where the true value lies, not in automation alone, but in augmentation that enhances both precision and care.

Building Intelligent Healthcare Systems That Operate in Reality

Designing physical AI systems for healthcare is fundamentally different from building traditional software. In controlled environments, systems can rely on predictable inputs and stable conditions. In healthcare, variability is constant. Patients differ in physiology, symptoms evolve over time, and environmental factors influence how systems behave. This means that intelligence cannot simply be accurate in ideal conditions. It must be resilient in real ones. Systems must function reliably despite noise, incomplete data, and unexpected scenarios.

This is where the concept of embodied intelligence becomes critical. Physical AI systems must integrate perception, reasoning, and action into a continuous loop. Sensors collect data from the environment, whether it is imaging systems, vital monitors, or embedded devices. That data is processed locally, interpreted in context, and translated into actionable insights. These insights then influence decisions, which in turn generate new data, creating a feedback loop that allows the system to adapt over time. This is not a one-time computation. It is an ongoing process that evolves with each interaction.

In practical terms, this means designing systems that are deeply integrated into workflows rather than layered on top of them. For example, in a surgical setting, Physical AI cannot operate as a separate tool that requires additional steps. It must function seamlessly within the procedure, providing guidance without interrupting the flow of work. In patient monitoring, systems must operate continuously without requiring constant manual intervention. The goal is not to add complexity, but to reduce it, making intelligence feel like a natural extension of the system rather than an external component.

Another important consideration is trust. In healthcare, trust is not optional. Clinicians need to understand how systems behave, and they need to rely on their outputs with confidence. This requires transparency in how decisions are made and consistency in how systems perform. Physical AI systems must be designed with clear feedback mechanisms, allowing users to interpret results and understand their implications. It is not enough for a system to be accurate. It must also be explainable in a way that aligns with clinical reasoning.

There is also a broader implication for how healthcare systems evolve. As Physical AI becomes more integrated, the role of infrastructure changes. Intelligence is no longer centralized. It is distributed across devices, systems, and environments. This creates a more flexible architecture where capabilities can be deployed closer to where they are needed. It allows for faster iteration, localized optimization, and more adaptive systems overall. Instead of relying on a single centralized model, intelligence becomes part of the fabric of the system itself.

What emerges from this shift is a new way of thinking about healthcare technology. It is no longer about building tools that assist from the outside. It is about creating systems that operate from within, continuously learning, adapting, and improving. Physical AI transforms healthcare from a collection of disconnected processes into an integrated, intelligent system. It aligns technology with the realities of care delivery, where timing, precision, and reliability are essential.

In the end, the success of physical AI in healthcare will not be measured by how advanced the models are, but by how seamlessly they integrate into real-world environments. It will be defined by whether they improve outcomes, reduce burden, and enhance the experience of both patients and providers. This is where technology moves beyond innovation and becomes impact.

Augmentation, Not Replacement

One of the most common fears around AI is that it will replace humans. This concern is understandable, but it is incomplete. The most powerful applications of physical AI are not about replacement. They are about augmentation. These systems extend human capability. They reduce repetitive work, improve accuracy, and enable faster decision making.

But they still rely on human judgment. The best systems are human-in-the-loop. A worker supported by AI becomes more effective. An engineer can detect failures earlier. A team can make better decisions. The goal is not to remove humans from the system. It is to make the system better.

Responsibility in a World Where AI Acts

As AI moves into the physical world, responsibility becomes central. When AI generates text, errors are often manageable. When AI interacts with the real world, errors can have real consequences. This raises important questions. How do we ensure fairness in these systems? How do we prevent bias from influencing decisions? Who is responsible when something goes wrong? These are not just technical challenges. They are human challenges. They require thoughtful design, governance, and accountability. We are not just building technology. We are shaping systems that affect how people live and work.

What This Means for the Future of Work

You do not need to be an AI expert to be part of this transformation. But you do need to understand it. The people who will thrive in this new world are not those who try to compete with AI. They are the ones who learn how to work with it. This means understanding what AI can and cannot do. It means developing judgment, not just technical skills. It means thinking in systems rather than tasks. The future will belong to those who can bridge technology and real-world application.

A Personal Reflection

When I think back to those early robotics projects, I realize how much they shaped my thinking. We were not just building machines. We were learning how systems behave in the real world. We were learning that even simple systems can produce unexpected outcomes. Today, we are building far more advanced systems, but the lesson remains the same: You cannot design intelligence in isolation. You must design it for the environment in which it will operate.

The Real AI Opportunity

AI is often described in terms of efficiency. Faster workflows, lower costs, increased productivity. While these benefits are real, they only scratch the surface. The deeper opportunity is about redesigning systems. It is about embedding intelligence into the infrastructure that powers industries. It is about creating systems that are more adaptive, more resilient, and more responsive. Physical AI represents a shift from intelligence as a tool to intelligence as part of the environment. It is a civilizational shift because it changes how we interact with the world around us.

Closing Thought

AI that generates text is impressive. AI that acts in the real world is transformative. As intelligence moves beyond the screen, we are entering a new phase of innovation. The question is not just what AI can do. The question is what we choose to build with it and how responsibly we choose to build it.

About the Author

Vinodhini Ravikumar is an AI leader, cofounder at Vault Voyage Limited, and engineering executive. She holds a master's degree in computer engineering and a doctorate in business administration with a specialization in AI. She has led large scale AI and distributed systems initiatives at Microsoft and is the founder of Mind Mosaic AI, focused on human-centered and inclusive AI systems. She is a Raptors Fellow, a member of the Forbes Technology Council, and an Edge AI researcher. Her work spans physical AI, multimodal systems, and real-world AI infrastructure with a focus on building responsible systems that augment human capability.

Email: vini@mindmosaicai.com

LinkedIn: https://www.linkedin.com/in/vinravik/

CHAPTER 21

FROM CAPACITY TO CONSCIENCE: THE REAL WORK OF ORGANIZATIONAL MATURITY

By Harvey Schinkal
Analytics Leader, Doctoral Candidate, Purdue University
Houston, Texas

A ship launched 50 years from now will overtake the one launched today.

The question is whether you understand why, and whether that changes what you do next.

—Andrew Kennedy's Wait Calculation (Paraphrased)

Several months ago, I was in a conversation with a colleague about AI adoption timelines, and he said something that stopped me mid-thought. We were discussing whether organizations should be moving faster to implement AI systems, and he offered a counterintuitive position: that waiting six months to start building actually lets other organizations do your research and development for you. The information becomes public. The mistakes get made by someone else. So you are not behind; you are subsidized.

I agreed with the surface argument but pushed back on what it implies. There is a formal version of this paradox in interstellar travel theory called the Wait Calculation, developed by Andrew Kennedy in 2006. The basic insight is that if you launch a ship to a distant star today with current propulsion technology, a ship launched fifty years from now with dramatically better technology will overtake it and arrive first. The math suggests you should keep waiting, because the next generation of capability will always be more efficient than the current one. But Kennedy's calculation also shows there is an optimal launch window. Delay past that point and you are not saving resources; you are forfeiting position.

AI adoption follows the same curve. The organizations that will have a sustained competitive advantage are not the ones that moved earliest or the ones that waited for the technology to stabilize. They are the ones that correctly identified the window, understood their own organizational constraints, and built the internal capacity to absorb successive iterations of the technology over time. The question most organizations are asking is: What AI tools should we implement? The question that actually determines outcomes is: How quickly can we learn, adapt, and integrate the next version of whatever comes after what we are building today?

That distinction between the tool and the organizational capacity to absorb the tool is the frame this chapter is built around. The ethics piece follows directly from it, because when organizations optimize for tool deployment without developing the capacity to understand what those tools are actually doing, they create the conditions for a specific class of harm that I will describe in detail. But the starting point is maturity, and most of the available thinking on that subject is missing the most important variable.

The Maturity Illusion

In my work advising organizations through analytics modernization, I have a recurring view of what AI adoption actually looks like from the inside of a mid-size enterprise. The pattern repeats across industries. An organization migrates from legacy reporting infrastructure to a modern cloud analytics stack. The business case is clear, the executive sponsorship is genuine, and the strategic intent is real. Then the project stalls, not because the technology failed, but because an upstream dependency, an ERP implementation, a data governance initiative, a workforce gap, was not ready to support it. The organization has the platform. It does not yet have the organizational conditions the platform requires.

That dependency is not a failure of planning. It is an accurate picture of what organizational AI and analytics maturity actually looks like in practice. The popular conversation treats maturity as a question of tooling: Do you have the platforms, do you have data scientists, are you using machine learning models in production? By those measures, many organizations in the middle of modernization programs look mature. By the measure that actually predicts long-term outcomes, which is the organization's capacity to absorb, govern, and build on successive technology iterations, they are doing the harder and less visible work of dependency mapping, architecture design, and change management that determines whether the tooling ever produces the value it promises.

I have observed three recurring failure modes when organizations mistake tool deployment for genuine AI maturity. The first is what I call the premature optimization trap: An organization acquires sophisticated AI tooling before it has the data foundations to support it, then discovers the tools are only as good as the inputs they receive. They have a Formula One car with a gravel road. The second is the governance gap: The organization deploys AI systems faster than it develops the internal expertise to understand what those systems are doing, which I will return to at length in the ethics section. The third is the velocity paradox: The organization treats AI maturity as a destination rather than a continuous capacity development process, which means they reach a plateau and stop investing in the learning infrastructure that would allow them to absorb the next wave.

The colleague I mentioned earlier put it well: Companies that go from zero to five on a maturity scale have an advantage not because they reached five, but because they got good at moving up the scale. When the next major capability shift arrives and resets the landscape, they are faster at rebuilding than competitors who treated their prior five as a permanent achievement.

A Framework for Organizational AI Maturity

Based on my work across enterprise analytics engagements, independent consulting practice, and ongoing doctoral research at Purdue University examining how ethics orientation in AI engineering can function as a structural do-no-harm principle during implementation, I have developed a working framework for organizational AI maturity that focuses on capacity dimensions rather than tool deployment status. The framework is organized around five interlocking capabilities that distinguish organizations that can sustainably absorb AI from those that cannot.

The first capability is data integrity. This is the most unglamorous dimension and the most frequently underinvested. It encompasses data quality, lineage, governance, and the organizational processes that ensure the data feeding AI systems reflects what it is supposed to reflect. A pattern I encounter consistently in consulting engagements is that legacy reporting environments carry a form of invisible institutional knowledge: the reports are maintained by people who know where the edge cases are, where the data is dirty, where the numbers need a judgment call before they are trusted. When organizations migrate to automated analytics platforms, that institutional knowledge does not transfer automatically. Data quality problems that were previously absorbed by human judgment become systematic errors at scale, and they are often harder to detect precisely because the new system looks authoritative.

The second capability is architectural modularity. Organizations that build AI systems as integrated monoliths find that replacing any component requires replacing the whole. Organizations that build modularly, with clearly defined interfaces between data ingestion, transformation, modeling, and presentation layers, can upgrade individual components as the technology evolves without rebuilding the entire stack. This is the technical expression of the Wait Calculation insight: The ability to absorb successive technology iterations depends on building for replaceability.

The third capability is interpretive governance, which is the organizational capacity to understand what AI systems are doing and why. This is distinct from technical audit capability, though it includes it. It encompasses the processes, roles, and decision-making structures that connect AI system behavior to organizational accountability. When an AI system produces an unexpected output, who is responsible for investigating it? Who has the authority to take the system offline while the investigation proceeds? These questions are often unanswered until an incident forces them.

The fourth capability is change absorption, which is the organization's ability to integrate new AI capabilities into existing workflows without disrupting the human processes those capabilities are supposed to support. This is fundamentally a change management capability, but it manifests in AI deployment as the difference between tools that get used and tools that get abandoned after the implementation project closes. Adoption rates, workflow integration depth, and the degree to which frontline users trust the system's outputs are the operational metrics that reveal whether this capability is present.

The fifth and most underappreciated capability is ethical architecture, which is the organization's ability to identify, evaluate, and mitigate the second-order effects of AI system behavior on people inside and outside the organization. This is where the ethics discussion begins, and it requires a more precise framework than most organizations currently apply.

When the System Works as Designed

In 2018, Amazon publicly discontinued an AI recruiting tool after discovering it was systematically disadvantaging female candidates. The story gets cited as a cautionary tale about AI bias, and that framing is not wrong, but it obscures the most important analytical insight. The Amazon system was not malfunctioning. It was doing exactly what it had been trained to do, which was identify candidates who looked like successful Amazon hires from historical data. The problem was that historical hiring patterns at Amazon, as at most major technology companies, had produced a workforce that skewed heavily male in engineering roles. The training data encoded that pattern. The algorithm learned it faithfully. The system worked as designed and produced outcomes that were systematically inequitable.

This distinction is the core of a framework I have been developing through my doctoral research, which examines how embedding ethics orientation directly into AI engineering practice can function as a structural do-no-harm principle rather than a post-deployment audit. I separate AI system failures into two categories. Direct alignment failures are cases where the system produces outputs that diverge from its design intent: the model misclassifies, the prediction is wrong, the pipeline produces errors. These are engineering failures, and they are addressed through technical audit, model validation, and system improvement. Social alignment failures are cases where the system performs exactly as designed but the design itself produces outcomes that undermine broader organizational or societal objectives. The hiring system is a social alignment failure. The algorithm worked. The objectives it was optimized toward were the problem.

Social alignment failures are structurally harder to detect and correct than direct alignment failures because there is no engineering signal that something is wrong. The model metrics look fine. Precision and recall are within acceptable ranges. The system is doing what it is supposed to do. The harm is happening in the space between what the system measures and what the organization actually values, and that gap is only visible if you are explicitly looking for it with the right measurement instruments.

The consequences extend well beyond hiring, which is why the framework has broader applicability than a single research domain. An AI diagnostic tool trained on medical data from a patient population that underrepresents certain demographic groups will produce diagnostic recommendations that are more reliable for patients who look like the training data. An AI credit-scoring model trained on historical lending patterns will reproduce whatever biases existed in those patterns, including the redlining and differential treatment that shaped credit histories across demographic groups for decades. In each case the system is technically functional. In each case the organizational and societal impact is harmful. In each case the failure mode is the same: The training data defines the system's understanding of what success looks like, and that definition contains embedded history.

The Training Data Problem

The most precise way to understand why social alignment failures occur at scale is to follow the data backward. AI systems learn objective functions from examples. In supervised learning, those examples are historical outcomes: who got hired and succeeded, who got a loan and repaid it, who received a diagnosis and responded well to treatment. The system learns to predict future outcomes by identifying patterns in past ones. This is a powerful capability, and it is also a mechanism that faithfully preserves whatever systematic biases existed in the historical process that generated those outcomes.

Consider what it means to train a hiring model on historical promotion and retention data from an organization with a homogeneous leadership pipeline. The model learns that certain educational institutions, certain career trajectories, certain communication styles, and certain professional network characteristics predict success. What it is actually learning is that success was historically concentrated among people with access to those institutions, those trajectories, those networks. A candidate from a historically excluded group who took a different path to equivalent competence looks like a pattern departure from the training distribution. The model will score that candidate lower, not because it is biased in the sense of having a prejudice, but because it has accurately learned a historical pattern that itself reflects systematic exclusion.

This is the training data problem in its most fundamental form. It is not primarily a problem of model architecture or algorithmic design, though both of those matter. It is a problem of what gets encoded in the data that defines the learning objective. When that data is a record of historically biased human decisions, the model becomes a bias amplification system with the additional property of appearing objective because the outputs are numerical rather than human.

The organizational implications are significant. Most companies deploying AI hiring systems do not have the engineering resources for meaningful algorithmic audit. They purchase tools from vendors, receive outputs in the form of candidate scores or recommendations, and make decisions on the basis of those outputs with limited visibility into the models generating them. The vendor's model was likely trained on data from other organizations, which means it is encoding industry-wide historical patterns rather than organization-specific ones. The bias is not just reproduced; it is averaged and laundered through the appearance of a third-party technical system.

The ethical architecture capability I described in the maturity framework is the organizational response to this problem. It requires that organizations deploying AI systems maintain the ability to audit those systems for social alignment failures, not just technical performance. It requires governance structures that assign accountability for second-order effects, not just first-order outputs. And it requires that organizations treat the training data question as a design decision with ethical implications rather than a technical input to be optimized for predictive accuracy.

Building Organizations That Can Absorb What's Coming

The AI governance conversation that is currently happening at the policy level is operating at a different timescale from the deployment decisions organizations are making today. The EU AI Act, EEOC guidance on algorithmic employment decisions, and emerging state-level regulation are all moving toward frameworks that will impose formal accountability requirements on organizations using AI in consequential decisions. Organizations that are building ethical architecture now are building toward a regulatory environment that does not fully exist yet but is clearly coming.

There is a practical argument for this that does not require any particular moral commitment. Organizations that treat algorithmic accountability as a compliance checkbox will build the minimum required infrastructure and no more, which means they will be structurally unprepared for the next governance requirement. Organizations that treat ethical architecture as a genuine operational capability will find that the same internal systems that satisfy today's requirements generate insight that improves system performance and stakeholder trust in ways that have direct business value.

In my consulting work, I have seen this dynamic operate in both directions. An organization that invested in data lineage and model interpretability documentation for governance purposes discovered, 18 months later, that the same documentation made debugging a production model failure dramatically faster than it would otherwise have been. The governance investment had an operational return that was not anticipated in the original business case. Conversely, an organization that skipped interpretability documentation to accelerate deployment found that when a model began producing anomalous outputs, the lack of documentation made root cause analysis so expensive that the organization eventually decommissioned the model entirely rather than fixing it. The shortcut created a longer path.

The capacity development model I have described throughout this chapter is the alternative to both of those failure modes. It requires accepting that AI maturity is not a project with a completion date. It is an ongoing organizational investment in the ability to understand, govern, adapt, and be accountable for increasingly complex systems over time. Organizations that accept that framing will make better decisions about where to invest and when, how to structure governance, and how to evaluate the real costs and benefits of AI deployment.

The Wait Calculation provides the organizing principle. There is no objectively correct moment to launch. There is an optimal window based on your current capabilities, your competitive position, and your honest assessment of your organization's capacity to absorb what you are building. Move too early without the foundations and you will spend enormous resources maintaining systems that cannot deliver their promise. Wait too long and you will cede the organizational learning that is the actual competitive asset. The calculation is not about the technology. It is about the organization.

What the Freight Demands

I want to return to where this chapter started, because the ethical and the practical are not separate arguments that happen to occupy adjacent sections. They are the same argument at different levels of resolution.

The organizations that will navigate the next decade of AI development well are the ones that understand what they are actually building when they deploy these systems. They are not building tools. They are building organizational capabilities, governance structures, feedback loops, and accountability mechanisms that will determine whether the outputs of those systems align with what the organization and the people it affects actually value. That is an organizational design problem as much as a technical one.

The social alignment failure framework I have been developing through my doctoral research is ultimately an argument that ethical performance and technical performance are not the same measurement, and that the only reliable way to align them is to build ethics into the engineering process itself rather than layering it on after deployment. Organizations that conflate the two will systematically underinvest in the latter until a failure makes the cost of that underinvestment visible. By then, the harm has been done. The candidates who did not get calls. The loans that were not issued. The people who received suboptimal diagnostic recommendations because they did not look like the training distribution.

The maturity framework argues that organizational capacity to absorb the next technology iteration is more valuable than any specific AI system currently in production. The ethics framework argues that the accountability infrastructure required to detect and correct social alignment failures is a non-negotiable component of that capacity. Together, they make the same claim: Organizations that build for the long game, with honest assessment of what their systems are doing and clear governance over the consequences, will outperform organizations that optimize for speed at the expense of understanding.

The ship is always safer at shore. But that is not what it was built for. The question worth asking is whether you understand the currents well enough to navigate the open water and whether the vessel is built to make corrections when you discover you were wrong about where you were going.

About the Author

Harvey Schinkal is an analytics leader and independent consultant with experience spanning enterprise analytics architecture, AI governance, and technology strategy across industrial manufacturing, utilities, and professional services. He is a doctoral candidate in Purdue University's Doctor of Technology program with an expected completion in 2027. His dissertation research investigates how ethics orientation embedded within AI engineering practice can serve as a structural do-no-harm principle during implementation, shifting ethical accountability upstream into design and build decisions rather than treating it as a compliance exercise after deployment. His original research framework distinguishes direct alignment failures, where AI systems diverge from their design intent, from social alignment failures, where systems perform as designed but produce inequitable outcomes at scale. He is a part-time instructor at Seattle Central College and operates an independent analytics consulting practice through Onca LLC.

Email: Harvey.Schinkal@oncaconsulting.com

LinkedIn: https://www.linkedin.com/in/harvey-schinkal/

CHAPTER 22

SUSTAINABLE ARTIFICIAL INTELLIGENCE (AI) FOR SUSTAINABLE BUSINESS SUCCESS

By Jasmin Jijina Sinkular, PhD
Technology & Sustainability Innovator & Consultant
Detroit, Michigan

Sustainable development is a fundamental break that's going to reshuffle the entire deck. There are companies today that are going to dominate in the future simply because they understand that.

—Francois-Henri Pinault

Put simply, sustainability means being successful today, without compromising the ability to be successful tomorrow. The concept of sustainability is particularly important for businesses that care about long-term or sustained success. No business entity can be successful in isolation. Instead, it must depend on a thriving interconnected ecosystem of businesses, economies, communities, natural resources and environmental factors to enable its sustained success. The transition to and maintenance of long-term sustainability is known as sustainable development.

AI holds tremendous potential for solving the most pressing problems, ranging from efficiently identifying chemical formulae for pharmaceuticals that cure longstanding diseases, to discovering molecular recipes for materials with desirable properties that solve product manufacturing challenges and beyond.

While AI's benefits to human life are unquestionable, there is urgency to resolve associated ethical, safety, and security concerns. In the hands of irresponsible or bad actors, AI can do much harm, whether it be by violating personal privacy, increasing the potential of human abuse or endangering lives due to software errors in autonomous vehicles or in weapons unchecked by human safeguards. These scenarios are unsustainable for business success.

Even in more commonplace scenarios, AI models require large amounts of energy to run and water to cool the hardware they run on. The hardware requires valuable natural resources to manufacture, eventually creating e-waste and cybersecurity risks if not processed circularly.

A thoughtful approach is essential to weighing AI business benefits against their fallout. Good judgment is needed to prioritize AI use cases, and AI systems (hardware and software) need efficient design to use natural resources and clean power judiciously, while leveraging circularity.

To account for mutual interdependence for business success, several frameworks have emerged to aid in understanding, measuring, and tracking the impacts, risks, and opportunities (IROs) that businesses present to each other, even as they compete. Example frameworks include the International Financial Reporting Standard (IFRS), the Global Reporting Initiative (GRI), and the European Sustainability Reporting Standard (ESRS).

Despite differences in frameworks across global regions, they aim to identify topics of importance to a business's stakeholders (i.e., investors, customers, suppliers, management) to prioritize these topics, related IROs, metrics, and targets.

Related to this, growing international concerns over the projected scarcity of shared planetary resources and repercussions of improper handling of waste (solid, liquid, and gaseous) have driven world leaders to define shared sets of economic goals to ensure sustained success for businesses and their supply and distribution chains. The Sustainable Development Goals (SDG) comprise one such framework, which consists of 17 goals that are a call to action to tackle natural disasters causing climate change, to preserve the earth's limited natural resources, improve health and education while promoting opportunities to spur a larger, more capable workforce, hence economic growth for sustained business prosperity.

The IFRS is another framework that identifies typical financial IROs for material topics relevant to each industry sector and defines metrics by which to measure the effectiveness of a business's transition to sustainable development.

These frameworks provide valuable guidance to business leaders seeking to minimize the many risks that plague their financial bottom lines and to deftly navigate their organizations to higher probabilities of sustained success.

AI's Sustainability: Its Bidirectional Impacts and Risks

AI dates back at least to the 1950s if not earlier, and there have been several periods since, when it became incredibly unpopular due to tall claims that did not match reality.

In the early 2000s, AI began to gain traction fueled by the introduction of graphical processing units (GPUs), better suited for parallel processing than traditional central processing units (CPUs), and the introduction of data driven approaches that replaced slower rules-based approaches. These advancements moved AI calculation times into the realm of realism, fueling its explosive growth.

Unfortunately, the accelerated performance of AI data center components comes at the expense of accelerated electricity and water consumption. While 2024 projections from the International Energy Agency (IEA) estimate that global electricity consumption could reach 945 terawatt-hours (TWh) by 2030, electricity demand specifically tied to AI data center servers is expected to rise even more rapidly (See Figure 1). That increase is being driven by the enormous power required to run advanced chips, servers, and cooling systems at scale.

As per the IEA's 2024 analysis of data centers, the typical breakdown of electricity demand for AI data center core components is about 60% for servers (high performance computing (HPC) components, e.g., GPUs, tensor processing units (TPUs), neural processing units (NPUs)); about 5% for storage (high speed and capacity storage components, e.g., nonvolatile memory express (NVMe), solid state devices (SSDs) with high bandwidth memory (HBM)); and about 5% for networking equipment (switches, routers, load balancers, fiber optics, co-packaged optics), IT equipment (UPS batteries, backup generators), and infrastructure (lighting, office equipment for staff), all typically crammed into space-saving configurations that make waste heat dissipation a challenge, requiring about 7% to 30% for cooling and environmental control.

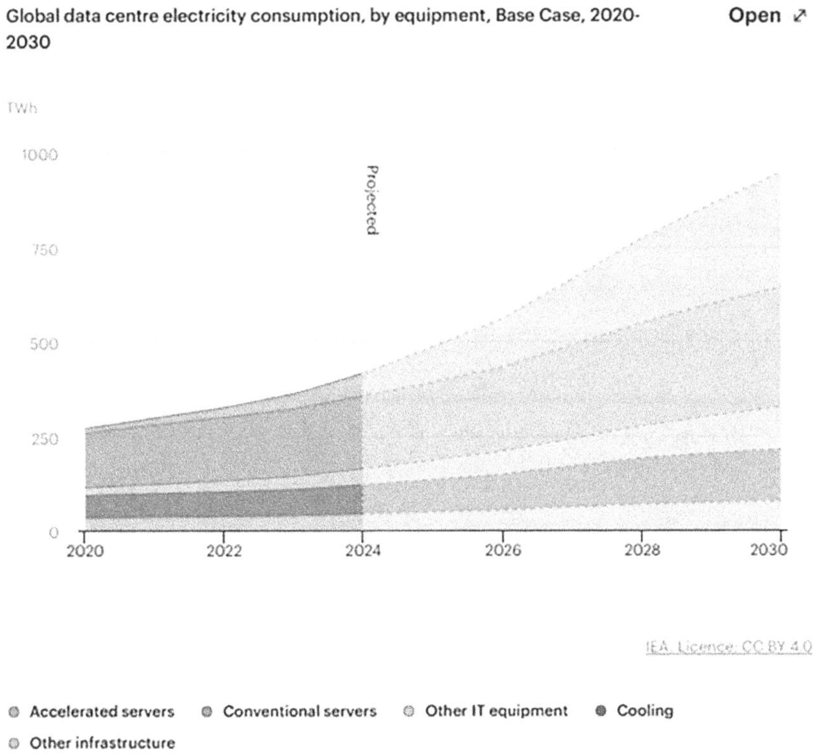

Figure 1: International Energy Agency (IEA) projections for annual electricity consumption by AI data center components.

Furthermore, as per 2026 statistics from All About AI (aaa), the leading AI engine ChatGPT4 used more than 50 gigawatt-hours (GWhs) to train its models and operationally uses about 625 megawatt-hours (MWh) daily for its 300 million weekly users. Additionally, US data centers consumed around 17 billion gallons of water in 2023 projected to increase four-fold to around 68 billion gallons of water by 2028.

Energy and water consumption constitute key IFRS material topics and metrics of interest to financial stakeholders of AI data centers as listed in Table 1 below:

#	Material Topic	Metric
1	Environmental Footprint of Hardware Infrastructure	Total energy consumed; %age grid electricity; %age renewable.
		Total water withdrawn; total water consumed; %age of each in regions with high or extremely high baseline water stress.
2	Data Security	# of data breaches; %age that are personal data breaches; # of users affected.
3	Data Privacy & Freedom of Expression	# of users whose information is used for secondary purposes.
		Total amount of monetary losses due to privacy legal proceedings.
		# of law enforcement requests for user info ad % disclosed.
4	IP Protection & Competitive Behavior	Total amount of monetary losses due to legal proceedings associated with anti-competitive regulations.

Table 1: Relevant material topics and metrics for IT data centers listed within the International Financial Reporting Standard (IFRS). Source: SELE Solutions, LLC, training material.

Other IFRS material topics include data security, data privacy, and IP protection. Data security breaches occur both during the operational lifetime of the AI data center components, as well as at end of life due to e-waste vulnerability. Properly implemented local circular processes can not only prevent such nefarious data breaches but also result in cost savings. Similarly, software and data processes can prevent financial risks from data privacy and IP protection breaches. In summary, a legal and ethical framework in business as in society is essential for a business's sustained success.

Next consider the Sustainable Development Goals (SDG) framework. Excess energy and water consumption by AI data centers causes negative impacts related to SDGs 6, 7, 11, 12, and 13. SDG 6 (clean water and sanitation) and 7 (affordable and clean energy) are especially notable when the introduction of AI data centers to an existing business ecosystem without proper water supply and transmission grid planning negatively impacts the availability, price, and reliability of energy and water for preexisting businesses in the area. This results in a failure of the interconnected ecosystem necessary for the long-term sustainability of AI data centers that rely on neighboring businesses to be their customers and to provide the components necessary for their sustained success.

This is ultimately disadvantageous to all businesses, including AI data centers themselves. When cities and communities fail to be sustainable (SDG 12: sustainable cities and communities), so do businesses within it. If the energy used by the AI data centers is unclean (e.g., coal, oil, and natural gas), or the materials used for data center construction (e.g., steel, cement) utilize unclean technologies, SDG 13 (climate action) is impacted negatively when excess greenhouse gas emissions (GHGs) cease to be in balance with the natural processes that absorb them, causing an increase in the frequency and intensity of natural disasters and in the associated economic toll. Excess GHGs and improper disposal of waste negatively impact SDG 14 (life below water) and SDG 15 (life on land), triggering employee health issues and biodiversity loss. Negative impacts related to SDGs 13, 14, and 15 cause negative economic impacts that disrupt business, impede supply and distribution chains, and destroy business facilities resulting in untold economic loss. (See Figures 2 and 3.)

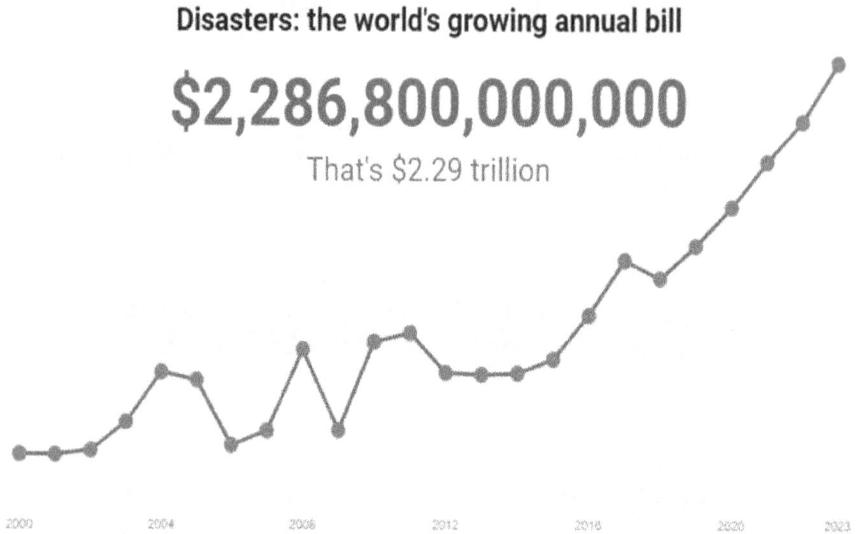

Figure 2: Growing annual cost of natural disasters from 2000–2023 exacerbated by climate change. Source: Global Assessment Report 2025, UNDRR

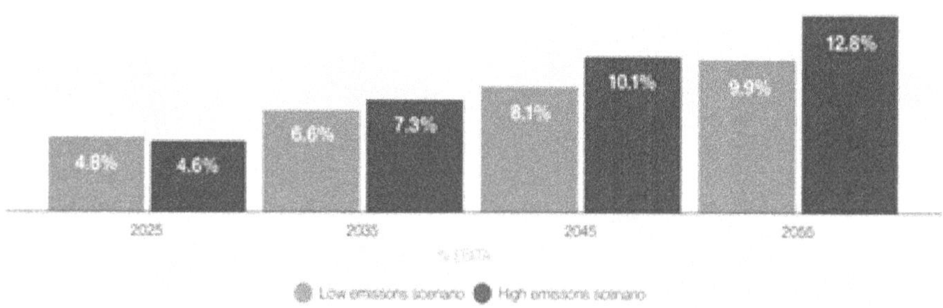

Figure 3: Fixed asset losses per year as a proportion of EBITA under low and high emissions scenario projections for 2025–2055. Source: S&P GlobalSustainable1, Accenture Analysis of 5043 companies.

Without adequately implemented circularity, the overuse and depletion of water and natural resources (e.g., copper for wires, silicon for semiconductors, lithium for battery storage, rare earth metals for electronic components and magnets) have negative impacts on SDG 8 (decent work and economic growth) as does mining in countries known to abuse workers' rights. This stunts economic growth that negatively impacts all businesses including the offending entity. SDG 9 (industry, innovation, and infrastructure) is impacted negatively by myopic, short-term strategies, SDG 10 (reduced inequalities) by unequal free market access to resources that cripple a vibrant business ecosystem, and SDG 11 (sustainable cities and communities) when the lack of a vibrant business ecosystem hinders sustained business success.

Just as with IFRS topics, the absence of circularity, improper mining, and e-waste result in negative impacts on SDG 14 and 15. Data privacy and security breaches result in the loss of reputation and trust that is essential for sustained business success.

Another negative impact of AI related to sustainability topics involves AI models incorporating and acting on implicit biases in the real world that find their way into model logic and reasoning. This can cause negative impacts related to SDG 5 (gender biases), SDG 10 (reduced inequalities), and SDG 4 (quality education), each of which diminishes the percentage of and training for the best and brightest minds available to businesses to spur sustained success.

Business leaders who wish to sustain success know the wisdom of the adage, "What goes around, comes around." The negative impacts of resource overuse on a timescale on which it cannot be renewed or circularly regained results in its scarcity leading to economic turmoil that poses financial risk. Disposing of waste (in solid, liquid, or gaseous forms) improperly, results in environmental impacts that increase financial risk for all businesses in the ecosystem. Treating workers or communities unfairly depletes trust and reputation without which investors, employees, and customers look to competitors who understand this. Business leaders who know how to steer their businesses clear of negative impacts on the economy, environment, and society can navigate their organizations towards sustained success.

Mutually Beneficial Opportunities for AI and Sustainability

Just as with other technologies, despite its risks, AI has tremendous potential to propel businesses and societies forward if balanced forethought and purposeful strategy are applied.

One such opportunity is self-serving, as it ensures the efficient use of energy and water for data centers. Facility and equipment sensors can monitor ambient conditions and feed data into AI models to optimize water usage and efficiently manage its circularity. Similarly, AI models with access to real-time information of surplus grid power or onsite renewable power can leverage battery storage efficiently. Similarly, statistical data that provides insights on power drawn by AI model batches can provide accurate real-time suggestions on temporal access and distribution of power for maximal energy efficiency of each software model and hardware component.

At its essence, sustainable development is about making a transition to better technologies, solutions, and processes that require fewer resources; using resources circularly or using more abundantly available renewable and clean resources; innovating to find better solutions that will continue to be economically viable and energy efficient while producing less harmful waste in solid, liquid, and gaseous forms.

The solution discovery process is unique to every problem. This often requires quickly sifting through commercially available technologies to determine what can best be adapted. If a commercially available solution does not exist, rapid innovation can result in new solutions and their optimization to reduce upfront investments and ongoing costs. While this valuable work requires human intelligence and guidance, AI's ability to efficiently search for potential solutions, using human expert guidance, and to provide best candidates for human-led discovery and commercialization can significantly shorten the solution discovery and R&D timeline.

In addition to rapid-fire innovation, sustainable development also requires data-centric measurement and tracking to key performance indicators (KPIs). AI does an excellent job of gathering big data from disparate sources and then inferring results such as natural resource availability predictions, supply chain disruption predictions due to climate induced natural disasters, GHG or water pollutant levels and their climate and natural resource impacts, economic risk models due to hurricanes, floods, wildfire hazards, and other hazards. A data-centric understanding of interconnectedness of human actions and their impacts on financial risk and opportunity is essential to maintaining sustainable success. AI systems are particularly adept at data intensive tasks as they can easily ingest large volumes of it, make sense of the underlying patterns and correlations, then create inferences that match or advance human understanding and interpretations.

A persisting challenge related to sustainability is the lack of awareness regarding its business implications. Sustainability education that caters to different audiences to build skills and awareness is crucial to achieving sustainable development and success. AI can easily adapt education curricula and styles to raise awareness amongst disparate audiences and help grow the necessary skills to tackle the challenges that lie ahead.

Lastly, as climate disasters increase in frequency and intensity, AI can play a pivotal role in analyzing big data to aid human experts in efficiently inferring reliable solutions for climate prediction, resilience and adaptation.

Experienced sustainability consultants like SELE Solutions, LLC, who are adept at problem-specific technology application can design and implement AI solutions that generate business value specific to each business and industry sector.

Sustainable AI: How to Transition to It and Maintain Success

Change is uncomfortable and requires action that businesses are typically unwilling to take unless compelled to do so. The need to achieve sustainable development can serve as the trigger and catalyst for positive change in the short and long term.

Each of the negative impacts and risks due to AI can be counteracted to produce an opportunity by transitioning to a sustainable technology or process. For example, the negative impacts of a large energy footprint due to AI can be replaced by the opportunity to increase energy efficiency via improved options for AI datacenter hardware, software, and operational processes and to use clean, renewable energy which has the lowest levelized cost of electricity (LCOE). As per a 2025 analysis by Lazard, the LCOE of utility scale solar ($38 to $78 per MWh) and wind ($37 to $86per MWh) is much lower than the LCOE of unclean or non-renewable sources: gas peaking ($149 to $251 per MWh), nuclear ($141 to $220 per MWh), coal ($71 to $173 per MWh), and gas combined cycle ($48 to $109 per MWh). This is to be expected as solar or wind energy is freely available despite being intermittent, is renewable, and does not require refueling once the power plant is constructed. Nuclear energy is clean but expensive and requires refueling and safe disposal of spent fuel. It also carries with it the baggage of national security and fuel safety.

Efficient energy use can also enable water conservancy. Innovative solutions to route or store waste heat for facility HVAC promote energy conservancy, as do thermally efficient geothermal solutions. When used for equipment cooling, geothermal solutions can optimize water usage. Thoughtfully designed data center architecture and prioritized use cases can further optimize energy and water consumption as can circularity.

Additionally, facility design and construction choices can improve the energy and GHG footprint of construction materials and of facility operations over their respective lifecycles. Equipment circularity can ensure precious resource conservation, hence avoid mining-related national security challenges and reduce costs.

In conclusion, a transition path to sustainable AI exists and is critical for success.

A Call to Action for Strategic Business Leaders

As monumental as the task seems, the transition to sustainable development is simply the no-nonsense path to acknowledging the reality of sustainability related to business risk and opportunity, then designing and implementing innovative transition solutions. Akin to financial reporting, sustainability reporting is essential for sustained business success.

So also, collaboration amongst business leaders via mutually beneficial agreements and regulations is essential for business ecosystem survival, which is a prerequisite for fair competition. It makes little sense for any given business to delay its transition to sustainable development in the hopes of making short-term, predatory advances at the expense of its competitors, as the consequences of doing so are unquestionably dire for all businesses.

For this reason, it is in the financial best interests of businesses to collaborate to create a baseline agreement on how to proceed with the sustainability transition in a savvy manner. Laissez faire capitalism cannot exist if the planetary and human resources that sustain it cease to exist.

About the Author

Dr. Jasmin Jijina Sinkular is a scientist by academic training, engineer by practical experience, and currently a sustainability consultant by intent. She cares deeply about helping her clients' businesses achieve sustainable success.

She earned a BS in physics from MIT and a PhD in physics from the University of Michigan, Ann Arbor, while conducting meaningful academic research for 11 years. To apply herself to creating business solutions, she then transitioned to industry where she worked both as an individual contributor and as a people leader for 20 years with world class technologists, engineers, data scientists, and industry thought leaders. She led super talented teams to zero defect commercial launches that spanned the entire product development cycle. For the sake of future generations, driven by her passion to move the world's businesses towards sustainable success, she founded her sustainability consulting practice, SELE Solutions, to leverage her training as a scientist, innovator, engineer, and implementer to deliver sustainable development. Through SELE Solutions, she helps business leaders understand why sustainability is important to business success, then helps them transition to sustainable development through identification, design, implementation, and tracking of customized technology solutions.

Email: info@selesolutions.com

Website: www.selesolutions.com

LinkedIn: https://www.linkedin.com/in/jasmin-jijina-sinkular-aa61544/

CHAPTER 23

HUMANS AND ROBOTS: AMPLIFYING LEADERSHIP IN THE AGE OF AI

By Sylvia Stocker
Global Robotics & AI Thought Leader
Zurich, Switzerland

Logic will get you from A to B. Imagination will take you everywhere.

—Albert Einstein

If your next employee never gets sick, never quits, learns faster than any human, and improves every single day, what becomes your role as a leader?

This is not a theoretical question anymore. This is physical AI—humanoid robots and AI joined in one machine—unfolding and acting in the real world. Think of places like logistics centres, production plants, hotel concierge, hospitals, and elderly homes. I see this as a technological transformation and at the same time a human transformation.

We are not facing a technological shift alone. We are facing a leadership moment of truth. Smart machines don't replace humanity; they amplify whatever leadership, mindset, and values already exist. The next era will not be defined by what machines are capable of. It will be defined by what leaders choose to amplify.

The Moment We Are In

We are living through a historical inflection point. Artificial intelligence is no longer confined to software. Robotics has stepped out of cages in factories and into human environments. What we are witnessing now is the rise of physical AI, machines that move, interact, assist, and increasingly collaborate with us. And yet, the biggest misunderstanding I encounter across boardrooms, stages, and strategy sessions is this: AI is still framed as disruption, as replacement, and often as a threat. However, that framing is both incomplete and dangerous.

Because AI does not fundamentally replace. AI amplifies and scales what is already there, meaning:

- Clarity or chaos
- Vision or confusion
- Strong leadership or its absence

This is the core principle that leaders must understand: Smart machines do not fix broken leadership. They expose it, and then they scale it.

The Stakes: Opportunity or Self-Inflicted Dystopia

There is a seductive narrative around AI, one that swings between utopia and dystopia. On one end, machines solve everything; on the other, they take everything from us. Both are wrong. AI is not the author of either future. Leadership is. We stand at a moment where AI and robotics can elevate human potential, freeing time, augmenting decision-making, enabling new forms of creativity, and addressing challenges in healthcare, aging societies, and global productivity. But the same technologies, placed in the hands of unclear, reactive, or fear-driven leadership, can just as easily lead to fragmentation, loss of trust, and systemic inefficiencies amplified at scale. I have seen both realities in deployment.

Personally, I see AI as a chance to evolve rather than a threat to survive. It is an opportunity for human progress in regards to robotics and intelligent systems and a clear warning: Without self-leadership, education, and courage, amplification turns fear into a self-fulfilling dystopia.

You Don't Speculate. You Operate

Across industries—hospitality, healthcare, pharma, corporate environments—at my company Arabesque we deploy robots and AI systems into real operations. Not pilot projects that live in innovation labs, but real environments.

And patterns emerge quickly.

-> The same robot

->The same software

->The same capabilities

Different outcomes. Why? Not because of technology. Because of leadership.

In one hotel, a service robot becomes a novelty, an expensive gadget that guests photograph once and then ignore. Staff sees it as a threat. Management never integrates it into workflows. The project quietly fades.

In another hotel, the same robot becomes an extension of the team. It handles repetitive delivery tasks, freeing staff to focus on guest experience. Employees are trained, involved, and empowered. The robot is not "the replacement"; it is part of a redesigned service model. Same technology. Different leadership.

Technology succeeds only when humans lead with clarity. This example reveals a simple truth: Organizations and individuals who thrive are not the most technical, but the most intentional.

The Core Truth

Let's land this clearly: AI will not decide our future. It amplifies our choices. Those who understand it, shape it, and work with it, benefit from AI. Those who wait, fear, deny, or delegate responsibility to the machine, lose. This is a call for education, not hype. It is likewise a call for leadership that scales through technology instead of being erased by it.

The question is no longer what AI can do. It is who we become when it does, our identity. I observe this every day, when people meet AI robots, especially humanoid robots, for the first time. Those who explore, ask questions, exchange ideas, and are open minded learn and adapt to the new circumstances the fastest. And that brings us to the five leadership shifts that define success in the age of AI.

Shift 1: Amplification

Define what you want to scale and what value it should create. Before implementing any AI and robotic system, leaders must answer a couple of simple questions:

- What exactly are we amplifying?
- What value does it create?
- What data do we need and do we have it?
- What are we willing to invest?
- Where is the human in-the-loop?

Because AI will take whatever exists—processes, decisions, biases, culture—and multiply it. If your organization is unclear, AI will scale confusion. If your data is flawed, AI will scale errors. If your culture lacks trust, AI will amplify resistance.

But if you are intentional, if your values, processes, and decisions are clear, AI becomes a force multiplier for excellence. One of the most overlooked leadership responsibilities today is not choosing the right tool. It is defining the right value, data, investment, and human-in-the-loop.

Shift 2: Responsibility

You cannot delegate leadership to machines. I'm observing a subtle but growing temptation in organizations: to outsource decision-making to algorithms.

- "Let the system decide."
- "The AI will optimise."
- "The data will guide us."

But here is the uncomfortable truth: Responsibility cannot be delegated to AI. AI can inform and suggest. It can even outperform humans in specific tasks. However, AI does not carry accountability. That will always remain human.

In one healthcare project, an AI-supported system was introduced to optimise patient flow. The technology worked. The data was accurate. The recommendations were clear. And yet, implementation stalled. Why?

Because leadership hesitated. They trusted the system, but not themselves to act on it. AI did not fail. Leadership did.

Shift 3: Mindset

From fear zone to self-leadership zone. Every leader I meet falls, at least initially, into one of two zones: the fear zone or the self-leadership zone.

In the fear zone, AI is perceived as a threat, something to control, delay, or avoid. Conversations revolve around risk, loss, and uncertainty.

In the self-leadership zone, AI becomes a tool for growth. Leaders ask different questions:

- What can this unlock?
- Where can we become better?
- How do we design this intentionally?

The shift between these zones is not driven by technology literacy alone. It is driven by mindset. And mindset is a leadership choice. I am not saying you should not consider the risks. I have witnessed moments in boardrooms where the energy shifts. A leader moves from skepticism to curiosity and exploration. This is where transformation begins.

Shift 4: Human Impact

Technology that unlocks humanity. Let me take you into a moment that has stayed with me. A true story.

In a care home in Germany, we introduced a humanoid robot, SoftBank's robot Pepper. The environment was cautious, staff were skeptical, and residents were curious but reserved. Among the residents was an elderly man who had not spoken to anyone for years. Not to staff, not to family, not to other residents. To Manfred, silence had become his world. Until something unexpected happened. One day, Manfred encountered the robot Pepper, a social humanoid robot that speaks to people and at the same time performs human gestures. So Pepper spoke to Manfred, and Manfred began to speak too! At first, quietly. Then slowly more engaging. Doctors called it remarkable and commissioned a study to the nearby university research lab. The staff called it emotional. I call it revealing!

Because it challenges a fundamental assumption: that technology reduces humanity. In this moment, it did the opposite. It unlocked it. This is what leaders must understand: The value of robotics and AI is not only in efficiency. It is in the human outcomes we rarely anticipate.

Shift 5: Reality vs. Hype

Grounding vision in execution. The AI space is full of bold claims:

- Fully autonomous systems
- Instant transformation
- Seamless integration

However, reality looks different!

- Integration can be messy.
- Adoption takes time.
- People need to be involved, trained, and heard.

Projects fail, not necessarily because the technology is incapable, but because expectations are misaligned with execution. Or, because values that the technology should bring are not defined. In one corporate deployment, a company invested heavily in AI tools without aligning teams or redefining processes. The result?

- Resistance, low usage of people
- High frustration in leaders
- Minimal impact for the organisation

Contrast that with organisations that start with a clear definition of value generation, investments allocated, data needed, and the defined role of the human-in-the-loop. Then start small, test, learn, iterate, and grow.

- A clear build-up
- No wait for perfection
- Instead build capability. And then capability compounds.

The Global Context: A Race Beyond Technology

In AI and in humanoid robotics, I see a race beyond technology. This is not only an organisational shift. It is a global one. The United States, China, and Europe, to mention a few big players, are shaping fundamentally different models of the future.

The US is building platforms, owning the intelligence layer, the compute, and the ecosystems others depend on. Key players include OpenAI, NVIDIA, Google, Anthropic, Tesla, and Boston Dynamics.

China is executing at industrial scale, turning AI and robotics into national infrastructure, deployed fast and widely on a global basis. Key players include Baidu, Alibaba, UBTECH Robotics, Unitree, Xiaomi, and BYD.

Europe holds depth in engineering, precision, and trust, currently strengthening its position. Key players include Mistral AI, ABB, KUKA, BMW, Neura Robotics, and Hexagon.

Who leads in AI and Robotics will influence productivity, economic power, and societal structures in the coming years. We are not just witnessing another wave of technology. We are entering another industrial revolution, but this time where machines don't just execute work; they understand it and act on it. The age of physical AI. But even at this scale, the same principle applies: Technology alone does not determine leadership. Intentionality does.

The Mirror Question

Let's return to where we started. If your next employee is a machine that improves every day, what becomes your role?

- Firstly, you do not compete with it. You lead with it.
- Your role changes. It becomes sharper, not smaller.
- You define direction, shape culture, take responsibility, and decide what gets amplified.

For most of us this requires personal growth.

The Mountain of Value and the Mountain of Possibility

You are sitting on a mountain of value. That is the message I often share with leaders. All of our experiences, our know-how, are valuable. The accumulated value over the years is your mountain of value. It is personal. And, it is growing every day. On the other hand, the mountain of possibility holds our opportunities, fantasies, ideas, imagination, and what could be. That part is unlimited and also personal. We humans have an endless source of ideas and imagination. It is a resource we can tap into at any time.

Now try this: With your imagination, and we all have it, picture these two mountains right beneath you. You are sitting on top of what we can call a mountain duet: the mountain of value and the mountain of possibility. You can tap into your expertise, know-how, ideas, and imagination at any time. This gives you strength and expands the opportunities you see.

Three Actions for Leaders

So what does this mean in practice? It means the following:

Action 1: Build AI understanding. If you lead, you must understand what you are scaling. Not at a technical depth but at a strategic one. You cannot outsource comprehension.

Action 2: Apply AI in real contexts. Clarity does not come from thinking. It comes from doing. Start where you are. Test, learn, and adjust.

Action 3: Strengthen your self-leadership. Technology will accelerate everything, including uncertainty. Your ability to lead yourself will define your ability to lead others.

Final Thought

Humans are made to grow, adapt, and learn. This is our DNA. We are individual beings with unique faces, bodies, voices, emotions, knowledge, experiences, and imagination. Once we accept that, we move beyond it. The question is no longer whether AI and robotics will shape our future. They already are. The real question is: What future are you choosing to amplify today?

About the Author

Sylvia Stocker, speaker, founder, and multiple award-winning CEO, has an impressive background in robotics, economics, and the performing arts. She exemplifies innovation and leadership in the field of robotics and artificial intelligence. As founder and CEO of Arabesque LLC, she has established herself not only as a technological visionary, but also as a tireless promoter of the human-centered perspective in robotics. Her unparalleled leadership and business acumen, combined with her large global client-base, make her a true source of inspiration and a role model for everyone in robotics and beyond. But Sylvia's influence extends far beyond her ventures at Arabesque. As a global speaker she shares her deep insights and experience to constantly push the boundaries of what is possible in robotics, AI, and leadership.

Book a keynote session with Sylvia: https://arabesque.ch/en/services/keynote/

LinkedIn: https://www.linkedin.com/in/sylviastocker/

News and updates on book chapter: https://arabesque.ch/en/book/

CHAPTER 24

SCALING LAWS: THE STRANGE PHYSICS OF INTELLIGENCE

By Sakina Syed
Microsoft IT Professional, AI Engineer, and Consultant
Toronto, Canada

What we observe is not nature itself, but nature exposed to our method of questioning.

—Werner Heisenberg

Intelligence as a Law of Nature

For much of our intellectual history, intelligence was treated as something built rather than discovered. Smart systems, whether minds, machines, or institutions, were assumed to arise from careful design, rare insight, and deliberate construction. Intelligence, in this view, was assembled through explicit rules, refined heuristics, and accumulated expertise. When a system failed, the solution seemed straightforward: improve the design.

Early artificial intelligence inherited this worldview. Researchers attempted to encode human reasoning directly into machines using logic trees, expert systems, and carefully engineered features meant to reflect how intelligence was thought to work. Progress was measured in cleverness and precision. Intelligence was something you built, component by component.

Then that assumption began to fail.

As machine learning models grew—not incrementally, but by orders of magnitude—their behavior changed. Performance no longer advanced in jagged leaps tied to conceptual breakthroughs. Instead, it improved smoothly with scale. Add more data, increase parameters, and expand compute, and error rates fell with striking regularity. These gains were consistent enough to be plotted, predicted, and extrapolated before training began.

This was deeply counterintuitive. It suggested that intelligence was not emerging primarily from clever design choices, but from scale itself.

The pattern resembled physics more than engineering. Gravity does not need to be reinvented for each planet; it follows a law. Heat does not require custom rules for every material; it obeys thermodynamics. Complex behavior emerges when underlying quantities cross thresholds. Intelligence, it appeared, might belong to the same class of phenomena.

As models continued to scale, they began exhibiting abilities no one had explicitly programmed: translation, abstraction, planning, reasoning. These capacities did not arise from new symbolic rules, but from accumulation—vast data, sufficient capacity, and prolonged training. Intelligence no longer looked like a brittle artifact assembled by hand. It looked like a property that emerges reliably in sufficiently large systems.

This realization reframes a foundational question: What happens when intelligence follows predictable laws rather than bespoke engineering? And what changes when intelligence becomes abundant rather than rare?

Why Bigger Models Behave in Predictable, Emergent Ways

To understand scaling laws, one must abandon a longstanding intuition: that intelligence depends on delicate, task-specific design. For decades, AI systems were narrow, fragile, and difficult to extend. Progress depended on expert tuning.

Scaling laws challenge that intuition. They show that intelligence improves smoothly with size, not specificity.

What Scaling Laws Reveal

Scaling laws describe a simple but profound relationship: As AI models grow in size, their performance improves in a smooth and predictable way. Double the parameters, expand the dataset, increase compute, and error rates reliably decrease along a mathematically describable curve. There are no sudden miracles, just steady improvement.

This relationship connects three variables:

- Parameters (the adjustable internal connections of a model)
- Data (the examples the model learns from)
- Compute (the processing power used during training)

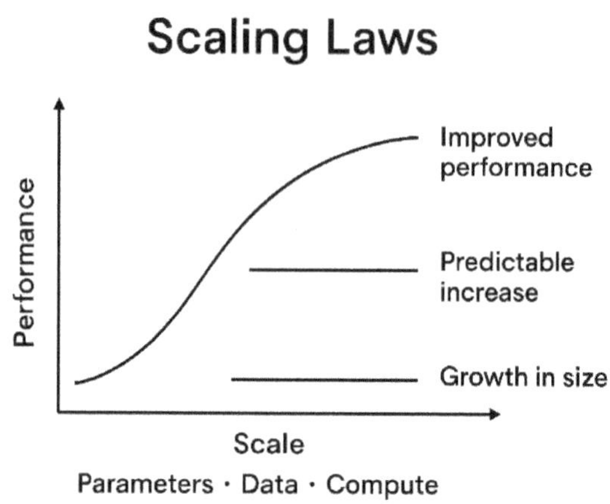

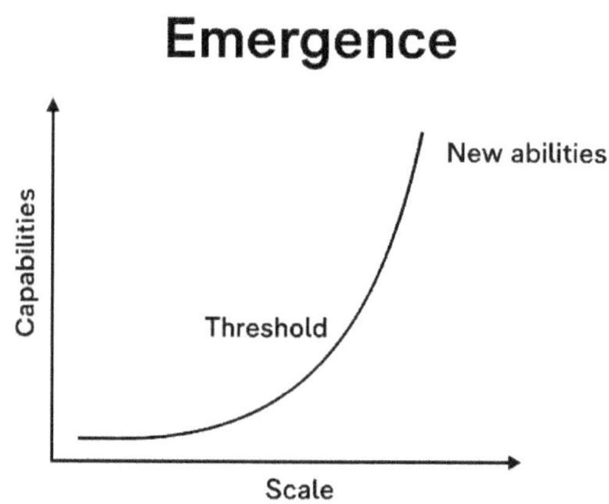

What makes scaling laws remarkable is not merely that bigger models perform better—that was already suspected—but that how much better they perform is predictable. Long before a model is trained, researchers can estimate its performance simply by knowing its size and data budget.

This discovery overturned decades of intuition. Earlier AI systems did not improve this way. Performance depended heavily on task-specific heuristics and handcrafted features. More data often led to diminishing or chaotic returns. Progress felt brittle.

Scaling laws revealed something radically different: Intelligence could be grown.

When Quantity Becomes Quality

As models cross certain size thresholds, they begin displaying capabilities absent at smaller scales. These are called "emergent behaviors," abilities that appear suddenly, even though no new mechanisms were added.

A model that struggles with basic language completion might, when scaled sufficiently, begin to translate languages it was never trained on explicitly. Another might suddenly solve multi-step reasoning problems, write functional code, or infer abstract relationships.

Crucially, these abilities cannot be "tuned into" smaller models. No amount of clever prompt engineering or fine-tuning can coax them into existence. They are properties of scale itself.

This mirrors phase transitions in nature. Water does not gradually become steam. It changes state once temperature crosses a threshold. Magnetism emerges when atomic alignment exceeds critical density. Life itself arises when chemical interactions become sufficiently complex.

In AI, quantity becomes quality.

The Predictability Paradox

To humans, emergent behaviors feel magical. A system that suddenly reasons or explains appears to have gained a mind overnight. But from the perspective of scaling laws, nothing discontinuous happened.

Performance curves are smooth. What changed is human perception. We notice intelligence when models cross thresholds meaningful to us. When they sound coherent, give useful answers, or solve problems we care about. The "aha moment" is psychological, not mathematical.

This leads to a paradox. Intelligence feels mysterious precisely because its growth follows laws rather than intention. The implications are unsettling. If intelligence is inevitable under certain conditions, then the question is no longer whether machines will become intelligent but how fast, how accessible, and to whom.

Why Architecture Matters Less Than Expected

Architectural innovations still matter. Better learning algorithms, attention mechanisms, and efficiency improvements all contribute to performance. But scaling laws imply diminishing returns to cleverness alone.

Across many model families, once scale dominates, differences in architecture blur. Models with very different internals begin to behave similarly once they are large enough. This convergence is deeply counterintuitive. It implies that intelligence is less about the specific wiring and more about the sheer expressive capacity of the system. Engineering becomes less about sculpting intelligence and more about enabling it.

Scale as a Path to Intelligence

The acceptance of scaling laws marked a turning point in AI research, a shift from theory-driven design to empirically discovered behavior.

The Pre-Scaling Era of AI: The Age of Handcrafted Intelligence

Early AI was built on explicit knowledge. Expert systems encoded rules provided by specialists. Symbolic reasoning engines manipulated logical statements. These systems succeeded in narrow domains and failed everywhere else.

They were brittle, expensive, and difficult to maintain. Each new domain required starting over. Progress slowed, frustrations mounted, and multiple "AI winters" followed.

The problem was not lack of intelligence among researchers. It was a mistaken premise: that intelligence could be fully articulated.

The Empirical Turn: "Just Train It Bigger"

The modern AI revolution began when researchers stopped insisting on understanding intelligence before building it. Instead, they trained models at unprecedented scale and observed what emerged. This marked a philosophical shift. Theory took a back seat to measurement. Intuition bowed to curves. Training runs became experiments, not deployments.

The mantra became simple: "Train it bigger and see what happens." Surprisingly often, what happened was intelligence.

The Three Pillars of Scale: Compute, Data, and Parameters as the New Ingredients

By this stage, the ingredients of scale are familiar. What matters now is not their definition, but their interaction. Capacity, experience, and computation reinforce one another in ways that make partial scaling ineffective. Increase one without the others and progress stalls. Models memorize without generalizing, data overwhelms insufficient capacity, or computation is wasted on systems unable to absorb it.

Scale is holistic or it fails. This explains why access to infrastructure matters as much as research talent and why AI progress increasingly resembles industrial capability.

From Optimization to Discovery

As AI models scale, something profound happens to the nature of research itself. The work begins to resemble physics more than programming. Instead of specifying behaviors in advance, researchers increasingly observe them. They measure phenomena—emergence, generalization, transfer—much as physicists study matter or astronomers chart new stars. Models are trained first; understanding follows later.

At first, this can feel unsettling. It suggests that intelligence can appear before we fully understand it. Scale produces behavior before explanation arrives. Capabilities emerge, and only afterward do we articulate why. But this is not a failure of science. It is a return to an older mode of discovery, one based on observation rather than prescription.

In this sense, modern AI learns much like a human child. No infant is born with explicit rules for language or physics. They are immersed in the world, and through exposure and interaction, structure emerges. Intelligence does not arise through instruction alone, but through sufficient data, experience, and capacity.

Large AI models operate in a similar way. They are not told what concepts mean or how reasoning should work. They adapt internally as they absorb information, and over time language coheres, abstraction forms, and reasoning becomes possible. The system surprises us not because it was explicitly taught, but because the conditions for intelligence were present.

Seen this way, the shift from optimization to discovery is not alarming but hopeful. It suggests that intelligence is not a fragile artifact requiring perfect design, but a robust phenomenon that emerges when the right ingredients align. Models are not fully engineered; they are grown.

The deeper implication is quietly optimistic. If intelligence can be discovered rather than prescribed, progress depends less on rare genius and more on patience, measurement, and alignment with underlying laws. Intelligence is not specified. It is discovered, and its growth remains generative and full of possibility.

Models That Improve Simply by Growing: Implications and Consequences

Once intelligence follows scaling laws, growth itself becomes a strategy.

Economic Consequences of Scale

Scaling laws favor those who can sustain scale. Compute is expensive, data accumulation compounds, and infrastructure advantages reinforce themselves over time. As a result, structural advantages concentrate among firms with the capital and resources to operate at extreme scale.

This creates a central paradox of the AI era. In principle, scaling laws democratize intelligence. Anyone can train a model using the same underlying methods. In practice, they centralize power, because reaching the most capable regimes requires resources few can afford.

A small number of organizations now sit at the apex of the intelligence curve, controlling performance regimes inaccessible to others. These positions become strategic choke points in the economy, not because of specific features, but because capacity itself becomes scarce. An arms race emerges, not for clever algorithms, but for compute, data, and sustained scale.

Yet concentration also generates a countervailing force. Rather than replicating foundation models, smaller companies can build atop them, leveraging open and prebuilt models that encapsulate the benefits of scale. By fine-tuning and integrating these models into specific domains, smaller players can move faster, focus on differentiated applications, and create value at the edges.

In this way, scale centralizes the production of intelligence while decentralizing its use. The economy reorganizes around intelligence as infrastructure—costly to generate but widely accessible to apply—allowing modestly resourced organizations to compete through creativity, speed, and domain insight rather than raw computational power.

Product Implications: Endless Improvement, Why AI Gets Better Over Time

Unlike traditional software, AI products improve without redesign. Conventional software advances through new features, rewrites, and versioned releases. Scaled AI, by contrast, improves primarily through training.

Each training run upgrades capability across many tasks at once. To users, this often feels like a sudden leap forward, even when the underlying progress was incremental and continuous.

Each new generation of models feels qualitatively different, not because engineers added new features, but because deeper intelligence unlocks entirely new capabilities. What appears as innovation is often simply growth crossing a threshold of usefulness. This creates a distinctive consumer experience: long periods of steady progress punctuated by moments of apparent transformation. Intelligence crosses utility thresholds, and users notice all at once.

The result is a reshaping of expectations. Intelligence becomes a moving baseline, and opportunity accrues to those who can recognize and act on these jumps as they happen.

Strategic and Societal Implications

Scaling laws reshape competition. Smaller players increasingly rely on foundation models rather than building their own. Nations prioritize access to compute as a strategic asset. Energy consumption becomes an AI bottleneck.

At the same time, reliance on scale carries risks. It encourages brute force over insight. It may obscure efficiency gains. And it raises environmental concerns as computation increases exponentially.

Future Opportunities in a World Governed by AI Scaling Laws

Scaling laws do more than explain why AI keeps improving. They redraw the map of opportunity. As intelligence becomes predictable, upgradable, and abundant, the sources of value shift. What once depended on rare expertise, large teams, or deep institutional memory increasingly depends on knowing where and how to apply intelligence that is no longer scarce.

This is the deeper consequence of scaling. It does not merely produce smarter machines; it changes who can act intelligently, at what scale, and with what leverage.

Opportunity Shift: From Building Intelligence to Applying It

In a scaling-law world, intelligence becomes a commodity layer. Like electricity or cloud computing, its production concentrates among providers capable of operating at extreme scale. Yet, as with electricity, the greatest value accrues not to those who generate it, but to those who use it effectively.

This creates a key decoupling. Domain expertise no longer requires deep AI engineering. What matters instead is problem selection, integration, and judgment. Competitive advantage shifts accordingly. Organizations stop asking, "Can we build a smarter model?" and ask, "Where can abundant intelligence generate disproportionate results?" The constraint moves from technical capability to imagination and context.

Organizational Opportunity: AI-Native Firms

As intelligence becomes continuously improving infrastructure, a new kind of organization emerges: the AI-native firm. These companies are built around the assumption that intelligence will keep getting cheaper, better, and more capable.

In such firms, AI functions as a universal collaborator, supporting research, strategy, analysis, and operations. This does not eliminate humans; it changes their role. Less effort goes into execution; more into synthesis, framing, and leadership.

The result is leverage. Small teams can now achieve outcomes that once required entire departments. Traditional metrics like headcount and budget matter less. In their place emerges intelligence leverage per person, the ability to amplify judgment with scalable cognition. Scaling laws thus favor both giants and insurgents: Large firms dominate infrastructure, while small teams dominate application.

Individual Opportunity: Cognitive Amplification

Scaling laws also reshape individuals. For the first time, a single person can access reasoning, writing, and analysis once reserved for institutions. This shifts what skill means. Execution becomes less scarce; judgment becomes more valuable. Knowing what to ask, what to discard, and what to pursue matters more than performing each step manually.

The rise of the solo enterprise reflects this shift: Individuals run businesses, research projects, or creative studios with minimal staff, supported by scalable intelligence. Education follows the same inversion. Learning becomes continuous and collaborative, not merely preparatory. The opportunity is not replacement, but amplification.

Societal Opportunity: Accelerated Discovery

At the societal level, scaling laws sharply reduce the cost of exploration. Many of humanity's hardest problems—disease, climate, materials, policy—are limited less by physical constraints than by cognitive ones.

When intelligence is abundant, hypotheses form faster, simulations scale, and iteration becomes cheap. This does not guarantee solutions, but it expands what is feasible. Problems once considered too complex or slow to study may become tractable. Intelligence shifts from a bottleneck into a multiplier.

Strategic Opportunity: Aligning with the Laws

History offers a final lesson. Organizations that resist underlying laws lose to those that align with them. In a world governed by scaling laws, advantage comes not from predicting AI's exact trajectory, but from positioning so improvement flows naturally through systems, markets, and people.

Intelligence is becoming inevitable. Opportunity belongs to those who design as if that inevitability is already here.

Limits, Plateaus, and Open Questions

Scaling laws are powerful, but they are not infinite. Physical constraints matter. Compute requires energy, hardware, and capital. Data is not endlessly fresh, and its quality varies. As models grow larger, diminishing returns may emerge, even if progress continues.

Many researchers suspect that scaling laws will bend rather than break. Intelligence may keep improving, but only by incorporating new paradigms: better data curation, multimodal learning, grounded interaction with the physical world, or even embodied systems. Scale alone may no longer be sufficient; it may need to be paired with deeper forms of experience.

Others raise a different concern: that scale can mask brittleness. Models may appear intelligent while lacking genuine understanding, producing fluent responses without robust internal grounding. Emergence without comprehension raises difficult questions about reliability, trust, and control, especially as systems are deployed in high-stakes settings.

The most honest conclusion is uncertainty. Scaling laws remain astonishing in what they explain but incomplete in what they guarantee. They have revealed a powerful path to intelligence, not a final theory of it.

Progress, then, is unlikely to stop abruptly. Instead, it may slow, shift, and change shape. The laws need not break to alter the trajectory of AI development. They may simply bend, guiding intelligence forward along new and less predictable paths.

Closing: When Intelligence Becomes Inevitable

Scaling laws transform our understanding of intelligence itself. Intelligence no longer appears as a singular achievement of genius or design, but as a predictable consequence of scale applied under optimization. When the conditions are right, intelligence emerges.

This realization reframes the AI opportunity at its core. Progress is no longer gated solely by brilliance or breakthrough ideas, but by access, access to compute, data, infrastructure, and increasingly, to shared models and tools that encapsulate prior advances. Intelligence becomes something that can be accumulated, managed, deployed, and contested, shifting from a rare trait to a strategic resource.

At the same time, this dynamic accelerates advancement across the field itself. Each generation of scaled models lowers the barrier for the next wave of innovation, enabling researchers, startups, and institutions to build atop existing intelligence rather than reinvent it. Opportunity multiplies not just at the frontier, but throughout the ecosystem, as progress compounds layer by layer.

When intelligence follows laws, growth begins to feel inevitable. The central question is no longer whether intelligence will expand, but who controls it, how it is governed, and whether its benefits are broadly shared or narrowly captured. As intelligence becomes infrastructure, responsibility shifts from inventors to stewards.

The strange physics of intelligence suggests something quietly radical: We may not be inventing minds so much as discovering the conditions under which they arise. And in a world where intelligence is inevitable, wisdom is no longer guaranteed, but it becomes essential.

About the Author

Sakina Syed is a dynamic senior data and AI consultant, AI engineer, and enthusiast with a stellar track record in the computer software industry, including notable tenures at tech giants like Microsoft. As a Microsoft AI engineer and Azure-certified professional, she excels in AI deployment, sales, management, teamwork, and leadership.

With a Bachelor of Science in Neuroscience and Mental Health Studies from the University of Toronto, Sakina has also enriched her expertise with courses in artificial intelligence, business management, and IT. Her passion for writing and traveling adds a unique dimension to her professional persona.

Sakina's enthusiasm for business and the application of business psychology to understand consumer needs is matched by her extensive seven-year experience in customer service. She is adept at interpreting consumer feedback and statistical data, enabling her to identify market requirements with precision. Her strong passion for team management and leadership drives her dedication to delivering exceptional customer experiences and fostering business success.

Email: sakina@youraiconsultant.ca

Website: www.youraiconsultant.ca

LinkedIn: https://www.linkedin.com/in/sakina-syed/

CHAPTER 25

QUANTUM COMPUTING AND THE FUTURE OF ECONOMIC DECISION-MAKING

By Vishruth Burra
DECA Award Winner; AI & Quantum Finance Researcher
Dublin, California

The difficulty lies not so much in developing new ideas as in escaping from old ones.

—John Maynard Keynes

Every civilization eventually confronts the limits of its tools. There was a time when economic systems were governed by intuition, experience, and simple arithmetic. As societies grew more complex, they demanded better models: statistics, calculus, simulations. Classical computing enabled an unprecedented leap in economic analysis, risk modeling, and financial engineering. It allowed us to process vast datasets, identify patterns, and make decisions at a scale that would have been unimaginable just decades ago.

Yet even the most advanced classical systems rely on simplification.

Markets are not linear. Economies are not isolated. Risk does not behave politely. Global systems—financial networks, supply chains, climate-linked economies—are deeply interconnected and dynamic. Feedback loops reinforce themselves. Small disruptions cascade into large consequences.

Classical computers approximate these systems because they must. They break complexity into manageable pieces, simulate scenarios within constraints, and rely on assumptions that smooth out the unpredictability of reality. These approximations have been useful, but they are, by design, incomplete.

Quantum computing challenges that constraint. It does not merely improve existing models. It invites us to reconsider whether our current frameworks for understanding economic systems are sufficient at all. It asks whether the limits we have accepted are truly inherent or simply reflections of the tools we have used.

A Different Way of Computing

Quantum computing is often described as "faster computing." This is misleading.

Quantum systems do not simply calculate more quickly; they explore possibility spaces differently. Through superposition and probabilistic states, they can evaluate multiple outcomes simultaneously, allowing them to address classes of problems that overwhelm classical machines.

Where classical systems follow defined paths, quantum systems operate across many potential paths at once. This enables them to capture relationships that are otherwise too complex, too interdependent, or too vast to model effectively.

This matters most not where answers are known, but where uncertainty dominates.

Finance, economics, and public policy are fields defined by uncertainty. They are shaped not just by data, but by human behavior, shifting incentives, and external shocks. Traditional models often struggle because they attempt to impose order on inherently uncertain systems.

Quantum computing does not eliminate uncertainty. It engages with it more directly. It provides a framework for exploring multiple futures simultaneously, rather than collapsing them prematurely into a single forecast. This represents a fundamental shift in how decisions are made. Instead of asking, "What will happen?" we begin to ask, "What could happen and with what likelihood?" That distinction is subtle, but transformative.

Rethinking Risk and Complexity

Modern financial systems are built on models that assume stability until proven otherwise. Stress tests simulate shocks, but only within predefined parameters. Correlations are estimated, not fully understood. These models work until they don't.

History has shown that the most significant disruptions often emerge from interactions that models failed to anticipate. The global financial crisis, supply chain collapses, and rapid market shifts all revealed the limitations of systems built on simplified assumptions.

Quantum approaches offer the possibility, still emerging, still imperfect, of modeling systems as they actually behave: interconnected, non-linear, and adaptive.

Instead of isolating variables, quantum models can incorporate interdependencies. Instead of assuming stability, they can explore instability as an inherent feature of the system. Instead of relying on fixed correlations, they can adapt to changing relationships in real time.

For corporations, this could reshape portfolio optimization, fraud detection, and long-term investment strategy. Risk could be understood not as a static metric, but as a dynamic landscape that evolves continuously.

For governments, it could improve economic forecasting, infrastructure planning, and policy simulation. Policies could be tested across a broader range of scenarios, revealing unintended consequences before they occur. More importantly, it forces a philosophical shift: from predicting single outcomes to understanding probability landscapes.

This shift requires humility. It acknowledges that certainty is often an illusion and that better decisions come not from eliminating uncertainty, but from navigating it more intelligently.

Power Without Wisdom Is Dangerous

History offers a warning. Better tools do not automatically lead to better decisions. They amplify intent. The same systems that optimize markets could destabilize them. The same models that improve efficiency could deepen inequality if applied without restraint.

Quantum computing magnifies this risk. Its power demands governance, transparency, and ethical foresight.

As decision-making becomes more complex, it may also become less understandable. Highly advanced models can produce outcomes that are difficult to interpret, even for experts. This creates a paradox: As our tools become more powerful, our ability to fully comprehend them may diminish. Without careful oversight, complexity becomes opacity, and opacity breeds mistrust.

This is particularly dangerous in economic systems, where trust is foundational. Markets function because participants believe in their fairness and stability. If decision-making processes become inscrutable, confidence can erode. The challenge, therefore, is not only to build powerful systems, but to ensure they remain interpretable, accountable, and aligned with human values.

Power must be paired with responsibility. Otherwise, progress risks becoming instability.

Scaling Beyond the Lab

Quantum computing will not replace classical systems overnight. Its integration will be gradual, hybrid, and uneven. In the near term, quantum systems will complement classical ones, addressing specific problems where their capabilities provide clear advantages. Over time, as the technology matures, their role may expand but always in coordination with existing systems.

The real challenge is not technical feasibility, but institutional readiness.

Enterprises and governments must learn how to:

- Interpret probabilistic outcomes responsibly
- Integrate quantum insights into human decision-making
- Avoid false certainty in highly complex systems

This requires new skills, new frameworks, and new forms of collaboration. Leaders must become comfortable with ambiguity. Decision-makers must learn to work with probabilities rather than definitive answers. Organizations must adapt their processes to incorporate insights that are inherently uncertain.

This is not a technology problem. It is a leadership problem. The institutions that succeed will not be those with the most advanced systems, but those that can integrate those systems thoughtfully into their decision-making processes.

Preparation begins now. The time to build understanding, governance, and capability is before the technology reaches full maturity.

Conclusion

Quantum computing invites us to confront an uncomfortable truth: The world has always been more complex than our models. For the first time, we may have tools capable of meeting that complexity head-on. But tools alone are not enough. Whether that leads to wiser decisions, or more sophisticated mistakes, depends not on qubits, but on judgment. It depends on our willingness to rethink assumptions, to embrace uncertainty, and to design systems that prioritize not just performance, but responsibility.

The future of economic decision-making will not be defined solely by computational power, but by how thoughtfully that power is applied.

In the end, quantum computing does not just expand what we can calculate. It expands what we must consider.

About the Author

Vishruth Burra is a student in the San Francisco Bay Area with a proven record in competitive business and technology innovation. An accomplished business competitor, he has earned top placements in regional and state-level competitions, including podium finishes in case study and roleplay events that assess real-world business strategy and communication skills. He has also been selected for Yale Young Global Scholars 2026 on the Solving Global Challenges track, a highly selective program recognizing outstanding student leaders from around the world. Motivated by his firsthand experience preparing for business competitions, Vishruth built an AI-powered platform that equips students with intelligent exam practice, AI-driven roleplay scenarios, a personal AI tutor, and detailed performance analytics, helping hundreds of students compete with greater confidence at every level.

LinkedIn: https://www.linkedin.com/in/vishruth-burra-3bbb08320/

CHAPTER 26

FROM CONNECTIVITY TO AUTONOMY: THE FUTURE OF HUMAN COLLABORATION AND AGENTIC AI

By Daniel Jonathan Valik
Technology & Product Executive; AI, Data Platforms, CX
Seattle, Washington

By Hardik Modi
Product Leader, AI, Data, & Marketing Tech Platforms
San Francisco, California

The best way to predict the future is to create it.
—Peter Drucker

The End of the Connectivity Era

In this chapter, we want to focus on the digital transformation being driven by agentic AI, and specifically, the profound impact it will have on the future role of communication services. We are going to take a deep dive into the foundational acronyms of our industry, Unified Communications as a Service (UCaaS), Contact Center as a Service (CCaaS), and Communications Platform as a Service (CPaaS) to understand exactly how they are coming together with the next wave of AI augmentation and agentic AI.

When you look back at the grand, sweeping history of human progress, it is ultimately a narrative about our tools and our evolving relationship with them. We built physical levers to lift heavier stones and construct cities, we engineered the printing press to democratize our ideas across entire continents, and we strung telegraph wires and telephone lines to carry our voices across the globe. For the better part of the last 40 years, the global technology sector has been almost exclusively obsessed with one singular concept: connectivity.

The entire goal was movement. We wanted to move a packet of data, a text string, a video frame, or a voice signal from point A to point B as fast, as securely, and as reliably as possible. We measured our success in milliseconds of latency and boasted about achieving "five nines" of uptime.

This relentless obsession with connectivity created trillion-dollar industries. Unified communications as a service (UCaaS) and contact center as a service (CCaaS) became the bedrock of the modern information age. We ripped out legacy, on-premises PBX hardware closets, replaced them with highly scalable cloud platforms, and gave every employee a unified client that worked seamlessly across their desktop and mobile devices. These systems were meticulously designed to connect us, to bridge physical distance, and to ensure that a corporate communication line never failed.

But right now, we are standing at the edge of a far more profound shift. We are moving beyond the era of simply connecting and decisively entering the era of autonomy.

This is the real AI opportunity. It is not just about incremental efficiency gains, trimming a few seconds off an average handle time, or basic cost reduction. It is a fundamental rewrite of how human collaboration and economic output are structured. The static tools we once used to merely "send" information are evolving into autonomous, goal-oriented systems that "act" on information. For business leaders, enterprise executives, and startup founders, this represents both an existential threat and an unprecedented, wealth-generating opportunity.

The Convergence: Blurring the Lines Between UCaaS, CCaaS, and CPaaS

To truly understand the magnitude of this transformation, we need to examine how enterprise communications have historically operated. For decades, the industry existed in rigid, isolated silos.

UCaaS enabled internal collaboration by providing the tools for chat, video calls, and file sharing strictly within the workforce. CCaaS, by contrast, powered external engagement, handling massive volumes of customer interactions across voice and digital channels while generating rich streams of customer, operational, and conversational data. These systems were procured by different buyers, managed by different IT administrators, and optimized separately, often by entirely different parts of the organization.

Despite their cloud-hosted sophistication, they were fundamentally passive utilities. Think of them exactly like the plumbing in your house or the electricity in your walls. They sat there completely silent and entirely useless until a human being actively engaged with them.

That utility mindset is now obsolete. The pipes have become intelligent.

To make this shift precise, we need to define exactly what we mean by agentic AI. Agentic AI systems are not the passive, simplistic chatbots of the 2010s that just responded to basic keyword prompts.

They are goal-directed systems capable of reasoning across multiple complex steps, maintaining context over a long period of time, and executing real actions across various enterprise systems to achieve defined outcomes. They do not simply assist human operators by surfacing a knowledge base article. They act, they adapt, and they continuously optimize their behavior within clearly defined guardrails.

This is also where communications platform as a service, or CPaaS, becomes incredibly critical. Historically, CPaaS was viewed strictly as a developer-centric layer. The raw APIs developers rented to embed SMS, voice, and video into any custom application. In other words, it abstracted complex telecom infrastructure into customer centric and consumable interfaces to build intelligent applications. In agentic AI-driven architecture, its role expands massively. CPaaS becomes the core orchestration layer that allows autonomous systems to execute physical actions across different channels and backend systems in real time while capturing and distributing the real-time interaction, event, and customer context data that agentic AI systems depend on to reason, plan, and act autonomously.

Think of CPaaS as the central nervous system of this new architecture. It provides programmable fabric through which AI agents can trigger complex workflows, integrate directly with global communications networks, and pull from enterprise data lakes to deliver outcomes. Without CPaaS, artificial intelligence remains a brain in a jar, smart but entirely disconnected from execution in the context of real time and cloud communications. With CPaaS, autonomy becomes fully operational, communicative, and scalable.

When agentic AI is introduced into this communication stack, the platform itself becomes an active participant in your business. It dissolves the hard boundary between internal collaboration (UCaaS) and external customer engagement (CCaaS), creating a single, unified system that operates seamlessly across both domains.

The modern contact center offers a brutally clear illustration of this transformation. Traditionally, human agents acted as human APIs. They sat in a swivel-chair environment, manually toggling between ten different browser tabs, retrieving information from a

CRM, reading a policy from a corporate intranet, and pasting the resolution into a chat window while the customer waited on hold.

In an agentic architecture, the platform assumes this entire role. It interprets the customer's raw intent, retrieves relevant data across all your backend systems, coordinates directly with internal knowledge bases, and executes the resolution in real time.

To ground this in a practical reality, consider a global e-commerce company experiencing sudden, massive disruption in its logistics network due to a weather event. In a traditional setup, this sparks a nightmare scenario. A massive surge of angry inbound customer inquiries, frantic manual escalations to Tier 3 support, and heavily delayed responses. The entire system chokes and is bound to limitations in real time.

In an agentic system, the supply chain disruption is detected automatically. The CCaaS AI instantly cross-references the impacted shipping manifests with the CRM. Affected customers are proactively notified via outbound CPaaS text messages, compensations or refunds are issued instantly based on predefined financial policies, and internal UCaaS channels are automatically updated so the executive team sees the resolution metrics in real time and all without a single manual human intervention. What was once a reactive, fragmented crisis becoming a proactive, unified, and smoothly handled event.

This is not simply automation. Traditional automation follows fragile, predefined rules and breaks the second conditions deviate from expectations. Agentic systems dynamically adapt to changing, chaotic conditions. They optimize toward a successful business outcome rather than rigidly executing a hard-coded process. This distinction is the absolute core of the disruption.

The implication is clear: The communications industry is aggressively shifting from enabling interactions to guaranteeing outcomes and business value.

The Economic Shift: Dismantling the Per-Seat Model

As the underlying technology of our enterprise evolves, we must innovate the economic models that support it. This is where the transition becomes highly disruptive for legacy vendors.

For the last two decades, the SaaS industry has relied almost entirely on per-seat pricing. Organizations paid 50 or a 100 dollars per user, per month, and vendors optimized their products to maximize daily active engagement and retention. This model made perfect sense for tools that required continuous human interaction. It was an economy based on human presence.

In an agentic environment, this alignment completely breaks down. When a single agentic AI system can comfortably perform the cognitive routing, data processing, and communication work previously handled by hundreds of human employees, charging an enterprise for a "seat" no longer reflects the value being delivered. It becomes a massive misalignment of financial incentives. Enterprises are exhausted by paying for "shelfware" licenses that sit idle or are only used for a fraction of their capability.

What emerges instead is an outcome-based financial model. Organizations are beginning to pay strictly for results rather than access. The primary unit of value shifts from presence to performance.

Consider an enterprise that deploys an autonomous system capable of handling both its massive external customer interactions and its internal IT service requests. Instead of signing a multi-million-dollar annual contract for thousands of agent licenses, the organization pays a dynamic rate based entirely on successful ticket resolutions, closed sales leads, or measurable customer retention.

This represents a violent structural shift from input-based economics, where human time and labor are the primary cost drivers, to output-based economics, where value is tied directly and exclusively to results. And that is exactly what we already see with business around the globe, a massive change towards increased business value and outcome.

This transition will be incredibly challenging for legacy providers built entirely on predictable, recurring per-seat revenue. They will need to fundamentally redesign their pricing strategies or face extinction. Simultaneously, enterprise organizations will need to develop new financial capabilities to measure and quantify the exact monetary value of their operational outcomes with mathematical precision. You cannot pay for outcomes if you don't know the value currency of it. The direction, however, is clear, and the shift is already actively underway.

The Evolution of the Information Worker: From Doer to Orchestrator

Beyond the shifting technology and economics, the most profound, deeply personal impact of the agentic AI revolution is its effect on the role of the everyday information worker.

For much of modern corporate history, knowledge work has been strictly defined by execution. Professionals handled the tasks, processed the endless spreadsheets, navigated the clunky legacy systems, and drafted the routine emails to achieve outcomes. We were the essential grease in the gears of the corporate machine.

As agentic AI systems absorb these routine responsibilities, the fundamental nature of white-collar work changes entirely. I will state this bluntly: If your primary job consists of mechanical information exchange or moving digital data from one folder to another, you are no longer the doer in the system. You are the bottleneck.

But this is not the end of the human worker. It marks the exciting, necessary emergence of the orchestrator.

Rather than measuring your daily productivity based on activity, such as the number of support calls you handled or the volume of emails you cleared out of your inbox before 5:00 p.m., productivity becomes a strict function of strategic clarity and control. The effectiveness of a professional will be determined entirely by their ability to define clear objectives, set rigorous constraints, and guide fleets of autonomous systems toward a desired result.

This transition requires an entirely different set of capabilities. Professionals must develop the ability to clearly articulate outcomes and translate abstract business goals into structured logic for AI systems. They must master prompt engineering, which isn't about writing Python code, but about deeply understanding how autonomous agents ingest data, sequence complex tasks, and execute across organizational boundaries without hallucinating.

Orchestrators must also excel at cross-functional synthesis. They need the ability to seamlessly connect the disparate dots between completely different business units, ensuring that while the AI handles massive data exchange, the overarching corporate strategy remains perfectly aligned.

In this context, the entire global economy shifts. It moves decisively away from a model based on human labor and steps into a model based entirely on human intent. The organizations that absolutely dominate this new era won't be the ones with the largest physical office buildings or the highest employee headcounts. They will be the lean, agile organizations equipped with the most skillful, visionary human orchestrators.

The Human Component: Empathy, Ethics, and Vision

As AI systems take on a massive, increasing share of daily execution, it begs a philosophical question: What exactly is left for us? The answer is simple. The uniquely human aspects of work become infinitely more significant.

AI is brilliant, almost terrifyingly effective at processing historical data, identifying complex patterns, and optimizing workflows within defined parameters. And from here, it only will become better and eventually super-intelligent. However, it does not possess genuine, radical empathy. It lacks deep moral and ethical judgment. And it is fundamentally incapable of defining a transformative, contrarian vision that departs from historical patterns, because it relies entirely on those historical patterns to function.

This creates a beautiful paradox. The more aggressively we automate routine, mechanical tasks, the more infinitely valuable these inherently human qualities become.

In a future environment heavily saturated with AI-generated content, synthetic voice agents, and perfectly optimized digital interactions, authentic human trust will rapidly become the absolute rarest and most highly sought-after commodity on the global market.

An agentic AI can seamlessly process a complex flight cancellation refund across your CCaaS and UCaaS platforms in two seconds, but it completely lacks the deep emotional judgment required to authentically console a weeping customer who is missing a family funeral because of that delay. It cannot feel the weight of a difficult decision.

The most successful leaders of the future will be those who lean more intentionally into their humanity. They will use agentic AI to aggressively remove operational friction, freeing their time for the work that machines cannot easily replicate: navigating ambiguity, having difficult and emotionally nuanced conversations, building meaningful relationships, mentoring the next generation of leaders, addressing complex ethical challenges, and defining long-term strategic direction.

The customer care industry provides an early glimpse into this transformation. AI is already handling a growing share of customer interactions that can be resolved through knowledge retrieval, process automation, and pattern recognition. This trend explains the rapid adoption of AI across the customer experience landscape. Yet the human role is not disappearing, it is moving higher up the value chain. The most complex situations, those requiring empathy, trust, judgment, negotiation, or creative problem solving, will continue to benefit from human involvement.

The same evolution is unfolding across virtually every form of collaboration. AI will help create content, manage schedules, identify relevant expertise, surface critical insights, and even recommend the right people to engage in order to move an initiative forward. Rather than replacing knowledge workers, agentic AI will increasingly act as an intelligent partner, augmenting human capabilities and allowing individuals to focus on the uniquely human aspects of innovation, leadership, and collaboration.

AI can easily determine exactly how to optimize a system for a given objective, and it can execute that optimization flawlessly at global scale. But it cannot determine whether that objective is morally or strategically appropriate. That responsibility remains firmly with humans. The distinction between "how" and "why" becomes the defining boundary of the next decade, and it is within the "why" that the enduring value of human judgment resides.

The Daily Reality of the Invisible Fabric

To understand the highly practical implications of this shift, it is useful to consider what a typical day feels like in an agentic environment. Let's ground this and imagine your daily routine on a completely average Tuesday morning, just a few short years from now.

You sit down at your desk to begin your workday. You do not open a cluttered, anxiety-inducing dashboard filled with flashing alerts and hundreds of unread messages. You do not log into a clunky CRM to manually update deal stages.

Instead, you interact with your control deck, a centralized, highly intelligent system that provides a synthesized view of business outcomes. The interface presents you with what already happened while you slept: Your autonomous agents operated continuously in the background, resolving global supply chain issues, updating complex product roadmaps based on customer feedback, and coordinating internal engineering workflows. At the same time, autonomous customer service agents continuously resolved routine service requests, processed account changes, coordinated appointments, and proactively engaged customers across voice and digital channels, escalating only the most complex or emotionally sensitive situations to human specialists.

Take a moment to internalize how that feels. You no longer feel a frantic, overwhelming sense of being constantly busy fighting operational fires. You have ultimate leverage.

Rather than reacting to basic problems as they arise, you focus strictly on higher-level activities. You spend your lunch hour building actual human rapport with a strategic partner. You spend your afternoon refining macro-strategy, debating the ethical guidelines of your company's new AI deployment, and designing entirely new customer experiences that require human creativity, judgment, and organizational alignment.

The operational burden is vastly reduced, but your level of strategic responsibility increases. Individuals move from managing digital tasks to shaping entirely new business systems. You operate with greater leverage, but also with greater accountability. AI provides a 360-degree operational view of key business metrics and drivers, proactively identifying opportunities for improvement, recommending automation, and continuously optimizing service operations across both human and AI agents.

This is not a distant, utopian science fiction narrative. It is the highly practical, logical extension of current technological trends across UCaaS, CCaaS, and CPaaS. Early versions of this exact reality are already actively emerging in enterprise environments today.

Conclusion: Embracing the Disruption

The transition to an agentic paradigm represents a massive, fundamental shift in how organizations operate, how we collaborate and communicate. It will reshape the communications industry, aggressively destroy legacy business models, and completely transform the nature of professional work itself.

To navigate this transition successfully, organizations must entirely rethink their approach to technology. Artificial intelligence can no longer be treated as a novel, shiny add-on feature or a basic chatbot layered onto existing legacy systems. It must be integrated as the foundational, intelligent fabric of the entire operating model.

This requires a complete redefinition of the product lifecycle. Traditional approaches, which focus on delivering discrete software features and periodic updates, are no longer sufficient. In an agentic environment, products must be designed as dynamic systems that continuously learn and evolve. This is fundamentally important as AI will increasingly service workers in the information, service, and knowledge sectors. This will provide us with an opportunity to shape the efficacy of human collaboration, across any area and use case. Every single interaction becomes part of a continuous feedback loop that informs future behavior, and success is measured not by software usage, but by the tangible business outcomes achieved.

This shift also heavily requires changes in organizational structure and corporate leadership. Companies must create environments that actively support orchestration, incentivize deep strategic thinking, and enable continuous adaptation. The role of leadership becomes less about micromanaging execution and much more about setting clear direction and ensuring strict alignment.

Communications services from CCaaS (customer intersection), UCaaS (knowledge and output workers), and CPaaS (data and channels) need to evolve rapidly with agentic AI in this new reality. Each of these communications service areas will soon provide out-of-the-box agentic AI agents and orchestration platforms (visual or process tooling for agentic AI), seamless integration into core business applications, and robust data services. In fact, deep connectivity into data services via MCP (model context protocol) or A2A (agent-to-agent) protocols already provides the necessary, underlying foundation for complex agentic AI workflows to function. And overall, data will be key as every data silo will be a unified integrated network to support these agentic AI workstreams.

We see forward-thinking software vendors already shifting their entire engineering efforts to provide agentic AI workflow and orchestration services. This allows agile enterprises to create, configure, and implement these intelligent services wherever business value exists, while fiercely providing the human in the loop with full, ultimate control and decision-making authority. This will ensure compliance, control, and sovereignty.

In other words, working in seamless collaboration with AI is simply the new normal. From here, we will see AI evolving rapidly from basic agentic task execution to true hyper-intelligence. Ultimately, the role of humans will continue to shift. Today we orchestrate these new AI services, but perhaps soon, AI will orchestrate the baseline operational systems for us, unlocking exponentially increasing value for humans to be purely creative and entirely focused on high-level, strategic decision making.

To summarize this profound shift, this transformation is not about replacing humans. It is about elevating them. As autonomous systems take over the heavy lifting of execution, humans are freed to focus on what they do best: defining purpose, exercising judgment, and creating meaning.

The era of connectivity is coming to an end. The era of autonomy has begun.

The question is no longer whether this technological transformation will happen. That ship has sailed, and it is already actively underway. The only question that matters now is whether you have the foresight and the courage to actively shape it.

It is time to start building.

About the Authors

Daniel Jonathan Valik is a product and technology executive, author, and global speaker with more than 26 years of experience driving innovation across cloud computing, communications services, data platforms, and artificial intelligence. Throughout his career, he has held leadership positions at organizations including AWS, Microsoft, Cisco, UJET, Vonage, RingCentral, VMware, and Huawei, helping build and scale products that serve millions of users worldwide.

Daniel's work has focused on the intersection of cloud communications, customer experience, data analytics, and AI, with a particular emphasis on the evolution of contact centers, unified communications, and communications platforms in the age of intelligent automation. Daniel has contributed to the development of industry-leading platforms and services spanning Microsoft 365, Skype, Microsoft Teams, Cisco Webex, Google AI, Amazon Connect Customer, and next-generation AI-powered customer engagement solutions.

A prolific author and contributor to more than a dozen technology book projects, Daniel is also a frequent speaker at global industry events, where he shares perspectives on product strategy, platform innovation, and the future of AI-powered business transformation. His insights are shaped by extensive international experience, including living and working in Europe, China, Hong Kong, Singapore, and the United States.

Daniel holds dual master's degrees in Strategic Management and continues his studies in artificial intelligence and machine learning through executive education programs at the Massachusetts Institute of Technology (MIT). He lives in the Seattle area with his family and remains passionate about technology, innovation, martial arts, and lifelong learning.

Email: Daniel@Valik.onmicrosoft.com

Website: https://www.amazon.com/author/danieljvalik

LinkedIn: https://www.linkedin.com/in/danielvalik/

Hardik Modi is a dynamic product management leader with a proven track record of launching impactful "zero to one" initiatives within large, matrixed organizations. Based in the San Francisco Bay Area, Hardik excels at recruiting and mentoring product managers, building high-performing product teams, and delivering cutting-edge AI/ML and data experiences for a diverse range of users, including end users, administrators, partners, and executives.

Hardik's expertise spans the entire product lifecycle, from strategic vision to successful execution. He spent over six years at Cisco, leading the Human Agent Insights and Analytics team. At Genesys he led predictive routing solutions using machine learning, generating significant revenue and evangelizing AI products. He also held key product roles at Lattice Engines, specializing in integrated data services and analytics engineering. As director of product management at ujet.cx, he is leading an AI-first approach in contact center as a service (CCaaS) platforms.

Academically, Hardik has continually invested in his growth through Stanford Continuing Studies, focusing on statistics for AI, machine learning, and data science, as well as product management. He holds certifications in generative AI and also is a recognized innovator, holding a patent for a "system and method for predictive routing based on a customer journey patience."

Email: hardike@gmail.com

LinkedIn: https://www.linkedin.com/in/hsmodi/

CHAPTER 27

THE MISSING INTELLIGENCE LAYER: HOW AI WILL HELP HUMANS NAVIGATE COMPLEX SYSTEMS

By Bobbie Wasserman, MBA
Founder, The Homeownership Collective; Civic Systems
Los Angeles, California

A wealth of information creates a poverty of attention.

–Herbert Simon

Modern life is full of experts, yet many people have never felt more on their own. We live in a world of specialists—financial advisors, insurance brokers, doctors, contractors, attorneys, consultants, and digital platforms for nearly everything—but when a real-life problem unfolds, the burden of putting the pieces together still falls on the individual. The result is a quiet but growing mismatch: Specialization has expanded, but meaningful support has not. What looks like a well-served society from the outside often feels very different from the inside. People are not simply searching for answers; they are trying to navigate systems that do not speak to one another and were never designed to.

That is the hidden structural truth. We do not have an information problem. We have a systems-navigation problem. The old model is broken because information is abundant, but integration is missing. Every domain protects its own lane, while the individual is left to absorb the consequences across all lanes: financial, logistical, legal, emotional, and strategic. In other words, the real burden is no longer just access to expertise. It is coordination across fragmented systems, under pressure, often with too little visibility into what comes next.

This is where the AI conversation becomes more interesting. Its most important role may not be automation alone, but something deeper: helping people move through complexity with greater clarity, context, and foresight.

The Complexity Problem: People Are Being Forced to Operate Across Silos

The deeper issue is not that modern systems lack expertise. It is that they are organized in ways that require ordinary people to function as the bridge between them. Finance, healthcare, insurance, housing, legal process, and public systems each operate with their own language, rules, incentives, timelines, and blind spots. In finance, a tax decision can alter long-term planning; in healthcare, a treatment choice can collide with coverage rules and billing realities; in insurance, a claim decision can affect not only immediate recovery, but future premiums, underwriting, and risk exposure. Within each domain, there may be highly trained professionals. But from the user's point of view, the experience is rarely coherent.

This is why the old story about complexity no longer holds. The problem is not simply that life has become more technical. It is that the burden of translation, sequencing, and coordination has shifted onto the person least equipped to carry it. People are expected to know which issue comes first, which expert matters most, what documentation to preserve, what risks are financial versus legal, and which "small" decision may create larger exposure later. The systems themselves are specialized, but the consequences are cumulative. Every silo hands off a piece; the individual is left holding the whole.

Modern life increasingly requires ordinary people to navigate expert systems without an integrating layer. That is not just inconvenient. It is becoming one of the defining pressures of contemporary life.

The result is a new form of vulnerability: not ignorance, but unmanaged complexity. That is where the next value layer begins to emerge. As systems become more specialized and more interdependent, value shifts to whatever can help people see the whole board, understand the sequence, and move through complexity with greater clarity.

The Missing Intelligence Layer

What, then, is the missing intelligence layer? It is not another service provider, another content library, or another isolated expert offering advice from within a single lane. It is the layer that helps people see the whole board. It connects the dots between systems that are typically experienced in pieces, and it helps individuals understand not only what is happening, but what comes next, what matters most, and where one decision may alter risk or opportunity somewhere else. In practical terms, it is the difference between receiving fragmented inputs and having a coherent frame for action.

This layer performs several functions that have become increasingly essential in modern life. It integrates information across domains rather than leaving people to stitch it together manually. It helps sequence decisions so that people can understand what must come first, what can wait, and what may create downstream consequences if handled out of order. It translates technical language and siloed advice into usable context. It makes trade-offs more visible. And perhaps most importantly, it helps surface the full chain of implications around a decision before those implications become expensive, stressful, or difficult to reverse.

That is the missing layer. It does not replace expertise, and it does not remove the need for human judgment. Instead, it creates a form of orientation that has been missing from the user experience for far too long. In a world defined by fragmented systems and interdependent consequences, the next great value layer is not more information. It is intelligence that helps people move across systems with clarity.

Why AI Changes the Equation

This is where AI changes the equation. Not because it "knows everything," and not because it eliminates the need for human judgment, but because it can perform a set of functions that fragmented systems handle poorly. AI can synthesize large amounts of information across domains, organize it into a usable structure, translate technical language into plain language, compare options side by side, and help model scenarios before a person takes action. In a world where complexity increasingly comes from disconnected systems rather than isolated problems, those capabilities matter enormously.

The real promise of AI is not simply speed. It is orientation. AI becomes most powerful not when it replaces judgment, but when it helps humans navigate complexity, sequence decisions, and surface consequences before mistakes become expensive. That makes it fundamentally different from a traditional search engine, a static database, or a narrow productivity tool. Search can return information. AI can help interpret context. Productivity tools can help execute tasks. AI can help people understand what task actually matters, what should happen first, and what risks may be forming just outside their field of view.

That is why AI is uniquely suited to become a navigation interface. It can sit across domains rather than inside just one of them. It can recognize that a financial decision may have insurance implications, that a maintenance issue may trigger legal exposure, or that a timing mistake in one area may increase cost in another. It can help people move from isolated answers to structured guidance. In other words, it can begin to fill the integrating layer that modern life has been missing.

Homeownership as the Proof Case

Homeownership is one of the clearest real-world examples of why the missing intelligence layer matters. For decades, the cultural story of homeownership was built around access: Save for the down payment, qualify for the mortgage, close the deal, build equity, and call the right people when something goes wrong. But that story no longer matches the lived reality of ownership. Homeownership is no longer a transaction. It is an ongoing system to be operated. The real burden of ownership is not just cost. It is coordination.

That is the rupture. The industry solved for the transaction and abandoned the operation. It built businesses around moments—buying, refinancing, insuring, repairing, selling—while homeowners live inside decision chains. The actual experience of owning a home is not defined by one milestone at a time. It is shaped by an ongoing series of interconnected judgments involving financing, insurance, maintenance, contractors, permits, taxes, governance, legal risk, and long-term capital allocation. A home is not just an asset. It is an asset wrapped in systems risk.

Consider what it now means to "manage" a home responsibly. The mortgage may be fixed, but insurance premiums may not be. Deferred maintenance may preserve short-term cash flow while increasing long-term exposure. A contractor decision may affect not only repair quality, but also warranty protections, code compliance, insurance documentation, and future resale disclosures. An HOA dispute may spill into legal and financial consequences. A tax decision tied to improvements or casualty loss may influence later planning. Even seemingly straightforward choices—replace now or wait, file a claim or absorb the cost, patch the issue or do the full fix—often carry consequences that cross multiple domains at once.

Water intrusion offers a vivid example. A homeowner notices staining on a ceiling or dampness near a wall after a storm. On the surface, it appears to be a repair issue. In reality, it may immediately become an insurance question, a remediation question, a contractor selection question, a permitting question, a documentation question, a financing question, and eventually a disclosure question. Should the homeowner file a claim, or will that create future underwriting risk? Should mitigation begin immediately, or will that complicate insurer review? Which contractor is qualified to diagnose the problem versus simply patch the symptom? Are permits required? Does the issue involve a common wall, roofline, drainage path, or HOA-controlled element? What records need to be preserved for a future dispute or eventual resale? If the homeowner lacks available cash, should the problem be financed, and if so, how does that affect the household's broader financial position?

None of these questions is irrational. All of them are normal. That is exactly the point. The problem is not that experts do not exist. Insurance professionals exist. Contractors exist. Remediation specialists exist. Attorneys, HOA managers, city departments, tax advisors, and inspectors all exist. The problem is that nobody is connecting the full chain from the homeowner's point of view. Every expert sees a lane. The homeowner lives the consequences across all lanes. That is why the homeowner has become the integration layer of last resort.

This is also why homeownership has matured into an operational reality even though the market still speaks to it in transactional language. The old support model assumes that if homeowners can locate the right expert at the right moment, they can manage the problem. But that assumption breaks down when the challenge is no longer one isolated task. The challenge is sequence. What comes first? What should be documented? Which decision creates downstream exposure? Where is the real trade-off between cost, timing, quality, risk, and future flexibility? Homeowners do not need more disconnected advice. They need a way to see the whole board.

Seen this way, homeownership is not just a useful example for understanding AI's future role. It is one of the strongest proof cases for it. The modern homeowner must increasingly act as project manager, risk officer, document keeper, contractor screener, financial planner, and dispute navigator, sometimes all at once. That is too much fragmentation for any ordinary person to manage well under pressure. Homeownership should feel like a triumph, not a trap. But for many people, what breaks them is not one catastrophic event. It is the cumulative drag of disconnected decisions, rising exposure, and no one helping them connect cause and effect. That is exactly the kind of environment in which an intelligence layer becomes not merely helpful, but necessary.

The Homeownership Collective is built around this model. This emerging model functions as a homeowner intelligence and navigation layer designed for the realities of modern ownership. Rather than treating a home issue as an isolated event, it helps people understand the full decision chain around that issue: what comes first, which systems are implicated, where documentation matters, what trade-offs are forming, and when a small choice may create larger downstream exposure. It provides guidance before a homeowner ever enters the silos of insurers, contractors, attorneys, HOA actors, or other specialists, so they can engage those experts from a more informed position and better understand how the information gathered in each silo fits into the larger operating picture of the home. By combining AI guidance, expert perspective, and shared intelligence from lived homeowner experience, it helps people move from scattered answers to structured decision-making.

AI as a Decision Partner

If the most important role of AI is to help people navigate complexity, then it is equally important to be clear about what that does not mean. It does not mean handing judgment over to a machine. It does not mean replacing expertise, accountability, or human discernment. And it does not mean that every decision can or should be automated. The more useful framing is this: The next great value layer is not more service providers. It is intelligence that helps people move across systems with clarity.

That is what makes AI powerful as a decision partner. It can improve human judgment by organizing complexity into something more visible and workable. It can make interdependence easier to see by showing how one decision may affect exposure in another domain. It can surface better questions before a person commits time, money, or trust in the wrong direction. And it can support readiness instead of reaction by helping people think through likely scenarios before they are operating under stress.

Used well, AI does not remove judgment. It gives judgment a better operating picture.

Who Will Thrive in the AI Era

The people who benefit most from AI will not simply be the ones who know how to open a chatbot window or generate faster outputs. They will be the ones who learn to use AI as a thinking partner. They will treat it as a navigation layer, a scenario-testing aid, and a way to reduce blind spots across systems they cannot possibly manage alone or entirely in their heads.

In practical terms, that means using AI not just to retrieve information, but to structure decisions. It means asking it to compare options, pressure-test assumptions, identify missing variables, and help clarify sequence and trade-offs. It means using AI to expand judgment rather than outsource it. The people who thrive in the AI era will not be those who surrender their thinking. They will be those who sharpen it by combining human discernment with structured intelligence.

The larger opportunity is not merely efficiency. It is greater clarity under complexity. In a world of fragmented systems, that becomes not just useful; it becomes a serious advantage.

The New Definition of Intelligence

As systems become more fragmented and the consequences of mis-sequenced decisions become more expensive, intelligence will increasingly mean the ability to navigate across domains, not just accumulate knowledge within one of them. That is why AI may become most valuable not as a replacement for human judgment, but as the missing layer of orientation, integration, and foresight people increasingly need.

The deeper shift is this: Value is moving toward whatever can help people see the whole board, understand the chain of consequences, and act with greater clarity inside systems that were never designed to make sense from the user's side. In that world, AI is not most powerful when it simply automates tasks. It is most powerful when it helps humans navigate complexity well. And that shift—from automation to navigation, from information to integration—may be the most important thing AI does for us.

About the Author

Bobbie Wasserman is the founder of The Homeownership Collective™ and Single Lady Estates®, with more than two decades of experience in strategic communications and public affairs helping organizations navigate complex, high-stakes environments. Her work focuses on the evolving realities of homeownership, the intersection of AI and human decision-making, and the need for better guidance across fragmented systems. She writes and speaks on housing, risk, governance, and the future of navigating modern life.

Email: hello@thehomeownershipcollective.com

Website: https://thehomeownershipcollective.com/

LinkedIn: https://www.linkedin.com/in/bobbiewasserman/

CHAPTER 28

AI IN LIFE SCIENCES: A JOURNEY OF TRANSFORMATION

By Cindy Weber
AI Solutions Leader
Medford, Massachusetts

If you have built castles in the air, your work need not be lost; that is where they should be. Now put the foundations under them.

—Henry David Thoreau, *Walden*

Artificial intelligence (AI) has the potential to deliver extraordinary impact across life sciences, but fulfilling that promise requires pragmatic innovation and broad adoption.

Within life sciences, and particularly the US pharmaceutical industry, AI is playing an increasingly critical role in improving patient care, supporting physicians, accelerating drug development, and reducing healthcare costs. Achieving sustained value from AI, however, is a journey rather than a destination.

Like any long journey, after early exuberance, the road must be faced with resilience, preparation, and adaptability. Otherwise, poor quality and eroded trust may provoke many to return to a familiar, comfortable routine.

How do we best navigate this journey together and achieve continuous transformation?

Applicable AI Methods, Including LLMs and Agentic AI

AI spans a wide range of approaches, from rule-based algorithms and traditional machine learning to the growing use of large language models (LLMs), as well as neural networks, generative adversarial networks (GANs), and other emerging methods. The recent growth explosion of commercially available foundational LLMs, such as Claude, Gemini, GPT, Llama, Mistral, and others, pretrained on vast amounts of public data, has motivated pharmaceutical companies to apply these models to help realize business improvements and efficiencies.

LLMs are adept at identifying data patterns, using deep learning methods, and are applied to do tasks like answering analytical questions, completing specific tasks based on more complex decision criteria, and predicting future events. The more traditional machine learning methods are increasingly combined-with or supplanted-by LLMs to achieve improved predictive results, such as, to identify if a patient may have a difficult-to-diagnose rare disease in order to help suggest diagnostic tests to physicians; to predict if patients may be nonresponsive to a medication or may require a treatment regimen change, to help physicians and patients to more effectively manage care and improve health outcomes; and to improve early tumor recognition in medical imaging, such as in mammograms and lung scans.

AI models are applied at the site of care, including within physician offices, by leveraging electronic medical record data in combination with AI-driven analytics. Pharmaceutical companies also apply these models to improve physician education and patient engagement using anonymized, HIPAA-compliant patient-level data. These models continue to improve through the integration of multiple AI methods and libraries, including the prediction of multiple future events and event sequences across diverse patient populations.

As more data is generated and captured, AI models improve in both accuracy and utility. The continued growth of electronic medical records, combined with increased automation and data enrichment across supply chains and medication dispensing, is amplifying the value of AI for pharmaceutical organizations.

Beyond the more traditional AI methods of rules-based and machine learning, the journey of applying LLMs, including autonomous and semi-autonomous AI agents, is getting accelerated attention from pharmaceutical companies. Though still in the early stages of their maturity curve and adoption, broader LLM-powered agentic ecosystems are being rapidly built to infuse AI into business processes. Acting independently or semi-independently from humans, these AI agents can reason, plan, and adapt as information is added and feedback is captured, and can act based on a changing context.

Pharmaceutical company agentic AI use cases, for example, can encompass:

- Conversational agentic AI interfaces that empower non-technical users to conduct analysis and explore complex questions across diverse structured and unstructured datasets, improving insight quality, reducing analytical effort, and supporting quicker, data-driven decisions.
- Autonomous agents used to detect data signals, including patient treatment discontinuation, insurance claim rejections, or product volumes falling below forecast; with agents quickly used to notify relevant teams and recommend targeted actions or interventions, or take specific actions autonomously.
- Agents that support the identification and summarization of peer-reviewed literature, and continuously monitor relevant market events, including, for example, the

publication of patient response studies for specific product classes, which may affect medication awareness and utilization.

Although company leadership may be impatient for rapid transformation, successful implementation and adoption of LLMs, agentic AI ecosystems, and related AI approaches requires a multi-step journey.

While this AI journey is complicated by the rapid pace of innovation and the expanding array of AI-enabled tools and services, a clear set of imperatives remains essential for navigating the inevitably bumpy path:

- The Guardrail Imperative
- The Data Imperative
- The Human Imperative
- The Process Imperative

The Guardrail Imperative

Strictly regulated for privacy and compliance, the pharmaceutical industry has guardrails in place for business operations that remain applicable as AI methods advance. Guardrails include the appropriate use of AI, and guidelines for AI in clinical practice and for when AI is considered a medical device. Governments have already taken actions to adapt privacy and compliance regulations for AI for the pharmaceutical and medical industries; and more will be needed as the AI methods evolve.

In addition, guardrails are important for quality and transparency, such as defined guardrails that report interpretation and use caveats related to AI responses, and AI that understands the boundaries and appropriate use of the underlying data domain. For example, an AI response may note areas of potential diagnostic concern, provide medical options and considerations, suggest additional areas of research, and quote from specifically cited medical guidelines, but

AI will not make a medical decision or treatment determination; instead, medical decision-making remains with the physician, patient, and healthcare proxy.

With the emergence of LLMs, a baseline of privacy and compliance regulation is critical for establishing trust in both the short and long term and for enabling broader adoption. However, regulations are not consistently defined across all industries. As consumer AI experiences have shown, when legal and regulatory guardrails are absent, or insufficient to guide and police AI behavior, response distortions and privacy breaches can proliferate rapidly.

The pharmaceutical industry, which is focused on applying AI to make medical and business-critical decisions, is, therefore, strongly motivated to deploy AI responsibly and transparently. Although the industry is data-driven, its stakeholders are human (patients and healthcare professionals) making trust and interpretability essential. Without transparency and standardized measures, people cannot reliably assess the accuracy of AI outputs or distinguish what is AI-generated content.

The guardrails of quality and appropriate use are an integral part of implementing AI along with the foundations of regulatory compliance, but managing an AI journey for accuracy, adoption, and transparency is not easy, which leads us to the data imperative.

The Data Imperative

LLMs and AI methods are, unsurprisingly, heavily dependent on data. Pharmaceutical company processes are very much data-driven, with that data including both unstructured data (documents, medical notes, primary research findings from interviews and surveys, etc.) and structured data (terabytes of data capturing pharmaceutical distribution, dispensing and use; educational outreach and awareness activity; etc.). In addition, AI can create synthetic data that can be leveraged for use cases such as training scenarios or marketing response simulations, expanding the data universe further.

While the "large language" of LLMs make these models better equipped to derive insights from unstructured data, the "small language" of row and column labels and data values of structured data provides insufficient information and context to power LLMs.

Even with the larger language of unstructured data, curation approaches for sourcing or grading data are important for company use cases reliant on more specialized subject areas, such as peer-reviewed medical research and reliable sources to track medical and market activity. AI models using unstructured data are also greatly improved by applying specialized ontologies and specialized LLM training for the subject area of interest. Trusted decision-making and business-critical processes demand specialized LLM implementations; however, AI cannot overcome poor study design in medical research, reinforcing the ongoing importance of human judgment combined with curated data and AI.

Trust is grounded in data. Historically, business intelligence platforms and a plethora of created reports have governed what users see through carefully curated interfaces and established processes. As companies move to LLM-driven ecosystems, this curation responsibility transitions from the presentation layer to the data layer.

Therefore, successful LLM deployments, especially those applied to structured pharmaceutical data, require data enrichment for metadata and context that is used by the LLM, including:

- Descriptions of the data fields and metrics
- Domain-specific dictionaries and ontologies
- Analysis sequencing instructions to answer different types of questions; for example, "Why is my product volume declining?" requires providing additional information to the LLM based on the subject area and available data.
- Grid and graph visualization creation or retrieval instructions
- Calculations to apply and when to apply them

- Explicit guidance on how content should be used and interpreted, with transparency around data caveats and domain boundaries

In addition to this data enrichment, datasets traditionally prepared for business intelligence applications must be restructured into agentic-ready data layers to support high-quality, fast LLM responses. This introduces a substantial volume of new metadata, context information, and re-engineered data to power subject-specific LLMs and agents, and is often the root cause of poor output quality and low adoption.

These foundational data layers depend on human expertise, which leads us to the human imperative for success.

The Human Imperative

People and their depth of expertise are foundational to the success of AI. Yet how many times are we pleased when AI returns an inadequate response? We feel our human value affirmed when that happens, which foregrounds an inevitable conflict: Iterations between humans and AI are needed for AI to learn, particularly as the use cases for AI get more specialized, but these iterations lead us to feel (often privately) that we are training our replacement. This is, of course, under-motivating. Companies need to face this conflict head on as part of their AI organizational change management programs, especially when the power dynamics of organizations often unintentionally encourage these fears to remain unspoken.

Though they have fears, people also feel a personal responsibility for their own expertise, so their integrity leads to the concern that non-experts will overly trust AI, when as experts they know AI can be imperfect and sometimes simply wrong, unless designed with care and attention. So, they fear being replaced but also fear the organization may unknowingly suffer when they are removed from the process. Despite this personal concern, often the company's experts are the ones who most readily embrace AI; they adopt more rapidly in many cases but want AI to be appropriately understood as human-enhancing rather than human-replacing.

The humanness of AI success and adoption requires approach changes that are of a type and scale not commonly implemented. Rarely has there been innovation that impacts everyone in the organization, all potentially in different ways and at the same time. The humanness of the diversity and emotionally-saturated range of responses is also rare from resistance and skepticism to embracing change; from feelings of AI marginalizing our humanity and integrity to enjoying the experimentation and adventure.

These dynamics necessitate an evolution in change management for AI rollouts and a ruthlessly honest approach to implementation; one that emphasizes transparency and quality over hype or accelerated claims of AI-driven impact. Equally important is deliberate consideration of how humans and AI work together, and which processes should responsibly transition to fully AI-driven execution.

These human considerations, along with the data imperatives, then lead us to consider how to shape successful processes for implementing AI, particularly for implementing LLMs and agentic AI into pharmaceutical company workflows.

The Process Imperative

Adoption is lagging (for now) for LLMs and agentic AI within pharmaceutical companies, primarily due to the data challenges, quality concerns, and the human dynamics. Top-down leadership goals for AI-driven cost savings and process changes, and the market need to reduce US healthcare costs, are fueling the organizational pressure for AI acceleration, but those goals are often misaligned with what is achievable and when.

Despite these pressures, success requires that we lean into humanness and be pragmatic about innovation. Rollouts should be intentionally phased and should assume iteration and ongoing management to enable those rollouts to be of high quality and more universally adopted. Companies need to realize, particularly due to rapid AI innovation, that the implementation journey never ends.

As companies navigate their AI journey, a strong focus on a set of core process assumptions is critical:

- While foundational LLMs may appear miraculous, deploying them in a business context requires domain-specific re-engineering and model retraining, with the experts in the organization playing a leading role in the design, context definition, training, and testing.
- Meaningful success is achieved beyond the initial implementation, through ongoing improvements in quality, transparency, and adoption, anchored by consistent and reproducible results.
- LLMs and agentic AI continue to depend on humans to ask the right questions and apply business acumen to the responses.
- Keeping humans in the loop for many AI-infused processes is important, except for discrete well-defined tasks that allow AI agents to act autonomously with high reliability.
- Effective organizational change management is essential and must be addressed both centrally and at the team and individual levels, recognizing varied and personalized responses to AI adoption.
- LLMs and agentic AI ecosystems need to be implemented wisely, for example:
- Distributed, orchestrated agentic AI ecosystems are required, since AI innovation is too rapid to realize competitiveness in isolation; thoughtful partnering for AI agents and specialized models will continue to be competitively important.
- Rapid change will continue for AI methods, among the LLM companies, and across the surrounding tooling and libraries, so adaptability over time is needed.

- Agent creation will become increasingly democratized allowing for increased rapid prototyping, and easier creation and deployment of agents by non-technical employees, with this proliferation needing to be managed to avoid waste and confusion.
- For quality, AI use cases must be intentional about determinism. The less deterministic, the higher the likelihood of misleading results, but the extrapolation and insight potential also increases (which is a value of LLMs); so, this balance must be adjusted for each use case and workflow.
- Set boundaries for each AI use case, so the LLM understands what it is unable to partially or fully answer or act upon, and be transparent to end users about that boundary.
- Be vigilant along the journey in looking for and managing sycophancy errors. Most LLMs and agents are set up to gather user feedback and adjust responses based on query history and context, but this can distort LLM responses over time if all feedback is treated equally.

Conclusion

No ambitious journey goes fully as planned, and this is particularly true for the AI journey, where innovation is rapid and human behavior is interwoven with the AI. Though AI methods infused into business processes will be increasingly critical for competitiveness and change, humans remain critical for this success. Successful companies will navigate the exuberance and disruption of AI and understand the foundational (and ongoing) hard work required across the imperatives of guardrails, data, humans, and process.

For pharmaceutical companies, this journey has the potential to be truly transformative by improving patient care while helping to reduce healthcare costs. As humans and patients, this transformation is ours, with AI as our powerful companion.

About the Author

Cindy Weber has over 20 years of experience in life sciences consulting and services, supporting pharma and biotech organizations with strategic and tactical commercial planning. She pursues innovation across AI, advanced analytics, technology, process optimization, and complex data, and is a business strategist known for driving transformational change by cutting through complexity and noise to focus on what truly matters and what is actionable.

A collaborative leader, Cindy believes excellence is achieved collectively and is devoted to working alongside smart people who are dedicated to a common goal; and as a long-time runner, she finds lightness and joy in movement and personal challenge. Cindy holds degrees from Bucknell University, the University of Delaware, and the University of Illinois Urbana-Champaign, and is currently a senior principal at IQVIA, where she continues to focus on AI-powered solutions and complex problem-solving.

Email: cindy@weber-pa.com

LinkedIn: www.linkedin.com/in/webercindy

CHAPTER 29

THE WISDOM GAP: WHY DISCERNMENT IS YOUR AI COMPETITIVE ADVANTAGE

By Christi Wharton
Mental Health and Professional Life Coach
Baton Rouge, Louisiana

Wisdom is the right use of knowledge. To know is not to be wise... But to know how to use knowledge is to have wisdom.

–Charles Spurgeon

The bottleneck in AI adoption isn't what most organizations think it is. It's not computing power, budget constraints, or access to the latest models. These are solvable problems. The real constraint is something more fundamental: We have vastly more capability than wisdom about how to use it.

This wisdom gap explains the paradox of our current moment. AI tools have never been more powerful or accessible, yet without thoughtful deployment, even the most sophisticated AI initiatives risk falling short of their potential. The problem isn't technical, the algorithms work, the systems process data, the outputs generate as designed. The failure is in judgment: automating processes that shouldn't be automated, deploying AI in contexts where it diminishes rather than enhances value, and moving so quickly that no one stops to ask whether the problem being solved is actually the problem that matters.

This pattern repeats across industries and applications. AI systems don't always behave as expected. They evolve, they surprise us, and they operate in contexts their designers never anticipated. What's missing is the discernment to navigate this uncertainty and recognize that technical performance and business value are not the same thing.

The wisdom gap manifests in the questions organizations fail to ask before they deploy. Instead of "Can AI do this?", the vital questions are: "Should AI do this?", "What are we optimizing for, and at what cost?", "Where does human judgment need to remain in the loop?", and "What happens to trust, relationships, and long-term value when we automate this interaction?"

The insight that separates organizations that thrive with AI from those that struggle: The winners aren't necessarily the most technically sophisticated. They're the ones who've developed the capacity to pause, evaluate, and choose wisely before they automate. They understand that in a world where AI capability is becoming commoditized, discernment becomes the scarce and valuable resource.

This reframes what the AI opportunity is. The disruption everyone talks about isn't just about technology, it's also about whether organizations will develop the wisdom to use technology well. Those who close the wisdom gap don't just avoid costly mistakes; they see opportunities others miss. They identify the processes where AI genuinely adds value rather than just cutting costs. They build trust with customers and employees by demonstrating thoughtfulness about when and how automation serves real human needs. They create competitive advantage not through technical prowess alone, but through the judgment to deploy that prowess strategically.

If discernment is the bottleneck, then the path to unlocking AI's potential isn't primarily about technical training or faster deployment cycles. It's about building organizational capacity to think clearly about what matters, to distinguish between efficiency and effectiveness, and to recognize that the ability to automate something doesn't necessarily mean automation is the right choice.

The wisdom gap represents both the greatest risk and the greatest opportunity in the AI age. Organizations racing to automate without developing discernment multiply their problems at machine speed. But those who cultivate wisdom about when and how to use AI position themselves to capture value that pure technical capability cannot deliver.

The question isn't whether your organization will adopt AI. It's whether you'll develop the discernment to adopt it wisely.

Discernment as Competitive Advantage

If discernment is the scarce resource, what does it look like when organizations possess it? And more importantly, how does it translate into competitive advantage?

Discernment in the context of AI isn't mystical or abstract; it's a practical capability that manifests in how organizations evaluate opportunities and make deployment decisions. It's the ability to distinguish between processes that genuinely benefit from automation and those that require human judgment. It's recognizing when AI recommendations might conflict with organizational values or customer needs. It's understanding second-order effects: If we automate this interaction, what happens next?

Organizations with discernment ask fundamentally different questions: "What problem are we actually trying to solve?" and "Will automation solve it or create new problems we haven't anticipated?"

They evaluate not just efficiency gains but impact on relationships, trust, and long-term value.

This thoughtfulness creates competitive advantage in three distinct ways.

First, discernment leads to better deployment decisions. Consider two healthcare organizations exploring AI for patient triage. The first sees an opportunity to reduce administrative burden, patients answer questions through an AI interface, which routes them to appropriate care. The system is technically sound and promises significant cost savings. They deploy quickly.

The second organization asks deeper questions: What happens to patients who struggle with technology? How do we handle cases where AI misses critical context? What's the patient experience when they're facing a health concern and encounter a machine instead of a human voice? They recognize that while AI can assist with triage, the initial human interaction is where trust and reassurance begin. They deploy AI as a support tool for staff, not a replacement for human contact.

The difference wasn't technical capability; it was discernment about what should be automated and what shouldn't.

This pattern applies across contexts. Discerning organizations invest in high-impact AI applications rather than chasing every trend. They avoid costly failures by recognizing early when automation might undermine rather than enhance value. The ROI isn't just in successful deployments; it's in the expensive mistakes they don't make.

Second, discernment creates trust and differentiation. In a market where everyone has access to similar AI tools, how an organization chooses to use them becomes a defining characteristic.

When organizations demonstrate wisdom about automation—thoughtfully choosing which interactions benefit from AI and preserving space for human judgment where it matters—customers and employees notice. The company becomes known not for having AI, but for using it thoughtfully. That reputation becomes a competitive moat in a world where poor AI experiences are common.

Third, discernment enables adaptive capacity. AI deployment isn't a one-time decision; it's an ongoing practice that requires course correction as context changes.

Organizations with discernment build cultures where questioning AI decisions is valued, not discouraged. When an AI system produces unexpected results or customer feedback suggests problems, they're positioned to respond quickly.

Consider a financial services firm that implements AI for loan approvals. A discerning organization doesn't just deploy and walk away. They monitor outcomes, audit decisions for bias, track customer satisfaction, and maintain human oversight for complex cases. When patterns emerge suggesting the AI might be missing important context or making decisions that conflict with their values, they adjust.

Adaptive capacity means building feedback loops that catch problems early and creating organizational permission to say, "This AI deployment isn't working the way we hoped, we need to rethink it." That permission is surprisingly rare. Many organizations feel committed to AI initiatives once deployed, even when evidence suggests they're creating more problems than value.

The competitive advantage of discernment compounds over time. Better deployment decisions lead to stronger customer and employee relationships. Trust becomes reputation. Adaptive capacity means learning from both successes and failures, building institutional wisdom about when AI adds value and when it doesn't.

This is why discernment isn't a constraint on innovation, it's what makes innovation sustainable. It's the difference between deploying AI because you can and deploying it because you should. That distinction determines whether artificial intelligence becomes a genuine competitive advantage or an expensive generator of problems that compound faster than you can solve them.

The Human Skills That Enable Discernment

Understanding discernment as competitive advantage raises an immediate question: What skills actually enable it? If the wisdom gap is the bottleneck and discernment is the unlock, how do individuals and organizations develop this capacity?

The answer isn't found in technical training alone. Discernment emerges from distinctly human capabilities, skills that become more valuable, not less, as AI grows more powerful. Three capabilities stand out as essential.

Critical Thinking in the Age of Instant Answers

AI excels at providing answers quickly. This creates a subtle but significant risk: the temptation to accept outputs at face value because they arrive with confidence and speed. Discernment requires the discipline to question and verify what AI produces, even when, especially when, it seems convincing.

Critical thinking in this context means asking, "What might this get wrong?" and "What's missing from this analysis?" It means evaluating whether AI is solving the right problem or simply optimizing for the wrong metrics. It means recognizing that AI systems reflect the data they're trained on, the objectives they're given, and the assumptions embedded in their design, all of which may be incomplete or flawed.

Organizations that cultivate critical thinking create cultures where questioning AI outputs is valued, not discouraged.

Ethical Reasoning Under Complexity

AI often presents decisions as optimization problems: Maximize this metric, minimize that cost, find the most efficient path. But many of the most important decisions aren't optimization problems, they're ethical dilemmas where efficiency conflicts with fairness, where speed conflicts with dignity, where what can be measured isn't what matters most.

Discernment requires the ability to recognize when AI recommendations, however technically sound, conflict with human values or organizational principles. It means understanding that "optimized" doesn't always mean "right," and that some things shouldn't be automated even when automation is possible.

Consider AI in hiring. A system might optimize for candidates who match patterns of past successful hires. But when those patterns go unexamined, optimization can perpetuate approaches that no longer serve the organization well. Ethical reasoning enables the discernment to ask: "Are we optimizing for the right outcomes?" and "Should we be replicating the past, or building toward a different future?

Ethical reasoning under complexity means developing comfort with nuance and ambiguity. It's the capacity to hold multiple considerations simultaneously—efficiency and fairness, innovation and caution, individual benefit and collective impact—and make wise choices even when the "optimal" answer isn't clear.

Systems Thinking

AI typically optimizes for defined objectives within bounded systems. Human discernment requires seeing beyond those boundaries to understand ripple effects, unintended consequences, and long-term implications.

Systems thinking means asking, "If we automate this, what happens next?" and tracing the answer through multiple levels. When customer service moves to AI chatbots, what happens to customer relationships? When that happens, what changes in customer lifetime value? When that changes, what shifts in competitive positioning?

It means recognizing connections AI might miss because they fall outside its training data or optimization parameters. An AI might recommend automating a particular workflow without understanding that the informal conversations happening during that workflow are where crucial knowledge transfer occurs. Remove the workflow, and you've inadvertently damaged organizational learning.

Organizations with strong systems thinking don't just evaluate AI deployments in isolation, they consider how each decision interacts with culture, relationships, processes, and long-term strategic goals.

These three capabilities—critical thinking, ethical reasoning, and systems thinking—aren't skills you learn once and check off a list. They're practices you develop continuously as AI evolves and as the contexts in which you're applying it change.

These are fundamentally human skills. They're not about competing with AI on its terms, processing speed, data analysis, pattern recognition. They're about bringing uniquely human judgment to bear on questions AI cannot answer: What matters? What's right? What are the consequences we care about beyond what can be easily measured?

Developing these capabilities at an individual level creates organizational capacity for discernment. When critical thinking, ethical reasoning, and systems thinking become embedded in how teams evaluate AI opportunities, the wisdom gap begins to close. This is how discernment scales from individual skill to competitive advantage.

The Choice Ahead

The AI age presents every organization and individual with a fundamental choice: Race to automate everything possible, or develop the discernment to automate wisely.

The first path is seductive. It promises efficiency, cost savings, and competitive advantage through speed. But this path creates problems that compound at machine speed.

The second path requires something harder: pausing to think before deploying, asking difficult questions about what should be automated rather than what can be, and building the human capabilities that enable wise choices. This path doesn't reject AI, it embraces the technology while maintaining clarity about when and how to use it well.

The paradox of this moment is that in a world where AI capability becomes commoditized, discernment becomes the differentiator. The organizations that thrive won't be those with the most sophisticated technology. They'll be the ones who've developed the wisdom to deploy that technology strategically, the critical thinking to evaluate outputs carefully, the ethical reasoning to navigate complexity, and the systems thinking to anticipate consequences.

Sustainable advantage comes not from automating fastest, but from automating wisest. The wisdom gap is real, and it's widening. But it's also closable, and closing it unlocks opportunities that pure technical capability cannot deliver.

The question isn't whether AI will transform your work. It's whether you'll develop the discernment to transform it well.

About the Author

Christi Wharton is a coach, author, and speaker who bridges neuroscience and Biblical wisdom to guide individuals navigating life's transitions with clarity, discernment, and confidence in their God-given calling. Christi holds a unique combination of certifications in mental health and brain health coaching, and AI for business. Her coaching blends Biblical wisdom and neuroscience with compassionate, practical guidance, equipping people to grow and embrace the life God has called them to live.

Christi is the author of *Wise with AI* and *God's Peace Code*, among others, and she creates life-transforming programs and coaching resources anchored in God's truth. Her work reflects a deep conviction that wisdom and discernment are essential tools for living and leading well in every season, including the age of artificial intelligence, and she serves those longing to live with clarity and purpose.

Email: christi@pathlightcoaching.com

Website: https://pathlightcoaching.com

LinkedIn: https://www.linkedin.com/in/christiwharton/

CHAPTER 30

GUARDRAILS AND GAS PEDALS: GOVERNING AI AS A BUSINESS DRIVER IN AN AGE OF RISK

By Eddie James Williams III
Head of Risk & Compliance (Deputy CISO)
Kansas City, Missouri

Until you were born, robots didn't dream, robots didn't desire unless we told them what to want.

—A.I. Artificial Intelligence (the movie)

Brothers and sisters, the decision has already been made. Artificial intelligence has already entered the room. Not the "future, hypothetical" room, or the lab. Not the conference stage where somebody in a designer blazer tells you the future arrived five minutes ago and will also, coincidentally, be available after the break at booth seven. I mean the boardroom. The budget meeting. The audit committee. Your city operations center. The hospital leadership call. That policy review you almost skipped. The procurement decision. Customer workflows. The public service queue. AI is already there, working away.

It is influencing decisions, accelerating output, shaping priorities, compressing timelines, and quietly, quietly altering how organizations think they understand risk. In some places it is formally approved and carefully managed. In way too many others it is used informally, inconsistently, and with more confidence than control. That tells us something important. AI is not just a *technology* story. It's a leadership, governance, and risk story. And probably more than anything else, it's a *decision* story.

For that reason alone, I can tell you that AI and cybersecurity as a whole aren't solely owned by IT, but organizational leadership.

Cybersecurity Was Always a Business Issue

As we all know, for years, organizations made the same mistake with cybersecurity. They treated it like a technical function, parked it inside IT, and only brought it to senior leadership after an incident, an audit finding, a regulator, or a customer forced the issue. Framing it like that was always too small to do anything useful (let alone fun) with.

Cybersecurity has never only been about technology. It has always been about driving trust, resilience, brand integrity, and revenue protection. In plainest terms, it has always been about enabling business.

AI makes all of that impossible to ignore. As soon as you develop a technology that can do things like influence judgment, summarize evidence, draft communications, recommend actions, prioritize work, or shape how humans understand a decision, it is no longer a back-office utility or app. Once that technology is born and is implemented it becomes part of the operating model. And anything that becomes part of the operating model belongs in the boardroom.

That is the first in a long list of truths that leaders are going to need to accept. Unfortunately, you might have to drag them to that realization. AI is a business capability with strategic upside and strategic consequence. If you handle it properly, you'll see speed, clarity, efficiency, and competitive advantage. If you handle it poorly, you create risk in the form of decision error at scale, which can lead to reputational damage, compliance exposure, and a false sense of maturity that will hurt worse than an honest lack of readiness.

The Opportunity Is Real, But So Is the Exposure

Now let me be equally clear about the opportunity. I am not anti-AI. I am not interested in fear for fear's sake, and I am not standing on the porch hollering at the future like it keyed my truck and insulted my mama (*and lord knows I'd fistfight a Cyberdyne Systems model T-800 for my mama*). AI is one of the most powerful opportunities organizations have had in a generation. It can reduce operational drag. It can improve throughput. It can bring to the surface patterns that humans like you or I may miss. AI and LLMs can accelerate research, analysis, documentation, and response. They can help leaders and teams reclaim time from repetitive work and redirect human effort toward judgment, strategy, and execution.

But...AI does not create wisdom. It doesn't create judgment, or accountability, and it damn sure doesn't create leadership. It just amplifies whatever we already have, and sometimes we don't have enough, or enough of the right stuff, focused, at least.

If your processes are disciplined, AI can scale disciplined execution. If your governance is weak, AI can scale weak governance. If your culture is thoughtful, AI can make thoughtful people faster. If your culture is sloppy, impatient, and addicted to shortcuts, AI can help you make bigger mistakes with cleaner formatting. We have been researching and putting together more and more AI security and governance case studies, and I'll tell you, if I had a nickel for every legal genAI, or LLM that made up cases and briefs wholesale, my next car would be a spaceship.

That is the part many leaders do not want to hear. AI doesn't remove organizational weakness. It reveals it, then multiplies it, then sells it back to you in your own voice.

The Illusion of Intelligence

The current hype cycle makes this a bit more dangerous because AI presents itself with confidence. It writes in full sentences and tidy summaries. It polishes its words enough to make people feel like something intelligent is happening, and humans are remarkably easy to impress when output arrives quickly and sounds halfway decently composed. This isn't a testament to the power of AI, it's a testament to the power of the human mind to anthropomorphize and commit the sin of pareidolia when we absolutely *cannot* afford it. Too many people hear confidence and assume competence; hell, we do that with other people too. Those same people work for organizations that see velocity and assume maturity. Too many executives see adoption and assume control. Now, if I had a nickel for every Michael Crichton sci-fi property that dealt with executives doing this exact thing, I'd have two nickels. Which isn't a lot, but it's weird that it happened twice (*and was a hit both times*).

That assumption is where risk begins.

AI is not intelligent in the way leadership needs it to be. It does not understand consequence. It does not carry fiduciary duty. It has no grasp of political context, institutional memory, public trust, or second-order business effects. None of that, unless humans deliberately build guardrails around its use and continue to verify the result. It has no idea that a recommendation which looks efficient on paper may create a fairness problem in practice, a regulatory problem in six months, a customer trust problem in a year, or a strategic problem the moment it reaches the wrong audience. Ask me one of these days how many security and business leaders I've had to redirect from the habit of blaming the toaster for burning the toast.

And yet organizations are increasingly letting AI influence, or even make decisions with real stakes. We are collectively starting to let the technology tail wag the organizational dog.

That is not innovation, That's just unmanaged decision risk.

What HUMINT Teaches That AI Cannot

My background includes human intelligence (HUMINT), and HUMINT teaches lessons that remain just as valuable in the AI era as they were before the first model learned to sound slick. In HUMINT, you do not take information at face value. You evaluate the source. You examine the context. You assess reliability, timing, incentive, and intent. You ask not only whether a piece of information is interesting, but whether it is credible, why it appears the way it does, what it leaves out, and what happens if you act on it.

Technology, especially AI does not remove that obligation. It increases it.

The difference is that human sources often have visible motives. AI often has invisible assumptions. That can make people trust it more, when in fact they should be challenging it more. Most leaders will challenge a vendor. They will challenge an analyst. They will challenge a consultant with a fancy deck and a weak point. Hand those same leaders a polished AI output with bullet points and a tone of certainty and suddenly half the room starts acting like the machine handed them tablets on a mountain, rather than an oddly formatted Word document.

How AI Actually Fails

Leaders also need to understand where AI actually fails. It rarely fails loudly. Traditional technology failure announces itself. Your systems go down, and your dashboards light up like Christmas trees. Users complain. Something breaks, let's fix it, simple. AI often fails quietly. It drifts. It nudges, and then it influences. A recommendation gets accepted because it looks reasonable. A summary gets trusted because it sounds complete. A prioritization engine starts shaping work because it is faster than human review. A chatbot gives a confident answer that is wrong in exactly the way a busy employee is least likely to question. A decision support tool becomes normalized before anyone has properly determined where human review has to remain non-negotiable.

Then one day, leadership discovers that a process they thought was efficient is actually biased, noncompliant, opaque, strategically foolish, or operationally unsafe.

By then, it is no longer a pilot problem. It is a business problem.

And don't get me wrong, I'm not picking on businesses. (Maybe I am a little bit). But other organizations using other types of AI tools, like the big ones folding proteins for medical research, can make mistakes too. You know what the difference is though? The review and testing process is built in. These big tools are used for research and as soon as a new variation or potential treatment for a prion disease is discovered and reported by the AI, they test the ever-loving hell out of it. It's in their nature, and how they "grew up" in scientific fields. I want us to apply the same type of, not skepticism, but rigor to other organizations when we use these tools. Simply because they are powerful and, more importantly, convincing enough for us to give them our control.

Start with the Decision, Not the Model

"OK, so what, Eddie? We've read this far and imagined you narrating this issue in your velvety baritone voice, but what do you expect us to do, just stop implementing AI tools, and miss the wave?"

Oh my goodness, thank you for the compliment, dear reader, but, no, we don't stop development, we don't stop innovation, and we don't stop business. But this is why leaders should stop beginning with the model and start beginning with the decision. Ask where AI can influence something consequential. Ask who is likely to rely on it, what data it touches, whether the process affects regulated information, customer outcomes, public trust, business continuity, money, or enterprise risk. Ask whether a wrong answer would simply be annoying or whether it would be material.

This is where you start governance and install your guardrails.

A Simple Model for Decision Risk

A simple way to think about AI risk is this. The technical chain matters, but the business exposure shows up at the moment a human decides what to trust and what to do next.

The critical control point is Human Interpretation, where trust is assigned, validation is either performed or skipped, and accountability becomes real.

Guardrails Before Gas Pedals

This is why I keep telling executive teams the same thing: Go as fast as you can to innovate, but put the guardrails in place to make your movement safe. Speed matters. Competitiveness matters. Nobody wants to be left behind while rivals automate, experiment, and learn. But speed without structure is not strategy. It's exposure wearing a sharp suit.

The organizations that win in this next phase of AI will not necessarily be the fastest movers. They will be the best-governed movers. They will know where AI is being used, who owns it, what decisions it can influence, what data it touches, what guardrails apply, how exceptions are handled, and when a human must intervene. They will distinguish between experimentation and production. They will understand the difference between a low-risk drafting use case and a high-impact decision support use case. They will not confuse novelty with readiness.

What Leadership Must Do Well

In practice, leadership needs to do several things well.

First, leaders need an enterprise view of where AI is actually being used. Not where they assume it is being used. Not only the tools that have gone through procurement. The real picture. Shadow use matters. Embedded vendor features matter, partially because everybody and their grandma has a tool that is now "powered by AI." Employees dropping sensitive content into public tools matter, because if there is a breach of a tool, there is a breach of your organization. Even more importantly, if anybody ever reads the terms-of-use agreements, whatever you put into the tool is now owned by the same group that owns the tool. Internal workarounds matter. If leadership does not know where AI is present, then leadership is not governing it.

Second, leaders need ownership that cannot dissolve into committee language until everybody is involved and nobody is responsible. This will kill momentum on your governance, and end up bringing even more risk. Cross-functional coordination is necessary. Security, legal, privacy, compliance, audit, procurement, operations, and data teams all have a part to play, so break barriers and build bridges to connect the groups that need to make decisions. But accountability still has to land somewhere. There must be a business owner, a risk owner, and an executive sponsor who can answer simple, uncomfortable questions when something goes wrong.

Third, organizations need those guardrails to be clear. What data can and cannot be used? Which use cases are acceptable? Which are prohibited? What constitutes high-impact use? What decisions require human validation? Who approves exceptions? What is monitored? What is reported to leadership and the board? If those answers are vague, you don't have governance, you have hope wrapped in a policy document.

Fourth, leadership needs reporting that speaks the language of the business. Boards do not need a technical lecture every quarter about model architectures and token limits unless those details materially affect risk. I've spanked the hands of CISOs that do that to this day. Board members don't care about that, they don't have the time, and they don't have the expertise, by design. We do. They need to know where AI is influencing important decisions, where trust is being placed, and whether the organization has a defensible operating model. We can't just hype an AI product or ecosystem. We have to have decision integrity.

Decision Integrity Is the Standard

Decision integrity is the standard that matters. Can the organization explain how an AI-influenced outcome was produced? Can leadership defend the process to a customer, a regulator, a judge, an auditor, a board, or the public? If the answer is no, then the organization does not have mature AI governance, regardless of how impressive the demo may have looked at the offsite.

The Human Risk Inside the AI Risk

There is also a workforce issue leaders need to take seriously. The more polished AI becomes, the easier it is for people to over-trust it. The danger is not only that AI may be wrong. The danger is that people stop challenging it. Once that happens, governance fails long before the headlines arrive. A polished answer becomes unearned authority. A shortcut becomes a substitute for thought. A team that should have been trained to verify starts learning to comply. That is the kind of shift that quietly degrades an organization from the inside.

Do Not Chase Tools. Build Judgment

That's also why I don't usually waste time teaching leaders how to use one specific AI tool for one narrow industry use case. Tools will change. Vendors will change. The names on the market map will change. Features will change every few weeks. What matters more is teaching leaders how to determine what matters for themselves. How to think about AI. How to evaluate it. How to govern it. How to separate signal from hype. How to tell whether a use case is safe, material, valuable, immature, or simply foolish dressed up as innovation.

Because there will always be a next phase.

The Teams That Will Actually Win

There is another dimension to this conversation that most organizations are not thinking about clearly enough. And I don't necessarily want to dive into it, but dip a pinkie toe in real quick before we get back to the rest.

AI is not just changing technology. It is changing what kind of teams win.

A while back, in a conversation with Tony, the executive director of CyberUp, we were talking about workforce development, and I kept coming back to a simple way of explaining it: You need bulldogs and brain surgeons. The bulldogs are the ones who execute. They move. They get after it. They don't overthink every step, but they are disciplined enough to follow structure, operate within guardrails, and keep momentum. They are the ones who make things happen. The brain surgeons are the ones who think. They slow things down when it matters. They question assumptions. They understand risk, consequence, and second-order effects. They are precise, deliberate, and accountable for outcomes.

Most organizations have one or the other. Very few have both working together effectively. AI makes that imbalance more dangerous. If you do not deliberately build teams that can have both, AI will expose that gap quickly.

Preparing for What Comes Next

The future of AI will absolutely not stop at better writing assistants and faster search. It's currently moving into robotics, autonomy, embedded decision support, healthcare workflows, citizen services, financial operations, software engineering, supply chains, identity systems, education, and defense. Who knows where it goes after that. The leaders who thrive will not be the ones who got lucky with a tool. They will be the ones who built organizations capable of adapting responsibly as the landscape changed.

And that is how individuals survive disruption too. They learn how to think, not just how to click. They learn what questions to ask and where accountability lives. They understand that routine work may compress, but judgment work rises in value. People who can connect technology to consequence, speed to safety, innovation to trust, and security to business outcomes will matter more in the years ahead. That is the crux of the opportunity inside the disruption.

Final Thought

AI is here. Between the boardroom and the government, between the vendor pitch and the citizen service, leaders need to know what they are dealing with in order to understand how to proceed.

I promise you the answer is not panic. Never has been, never will be. Panic will paralyze you.

You need disciplined movement.

Move forward. Move boldly. But do it with structure.

Guardrails before gas pedals. Drive fast, but wear your seatbelt.

That's good stewardship and respect for risk. And stewardship will separate the winners from the losers in the AI era.

The winners will align AI to trust, compliance, resilience, and revenue. They will move fast, but with clarity. They will understand that security is not friction against the business. It is part of how the business earns the right to grow. They will treat governance not as a brake, but as a way to scale safely, credibly, and with the confidence of customers, citizens, regulators, and investors.

The losers will confuse motion for maturity. They will chase output without ownership. They will allow AI to influence decisions they cannot explain, defend, or control. They will discover too late that confidence is not intelligence and speed is not wisdom. They will learn the hard way that a machine can help you move faster without helping you move better.

AI does not own outcomes. You can neither chide nor congratulate the toaster for how the bread turns out.

Leadership does. That will remain true no matter how advanced the technology becomes. And regardless of the "next big thing" that comes along.

So the question is not whether your organization will use AI. Most already are. The real question is whether you are building an organization ready to take advantage of what AI can do today while preparing responsibly for what it will do tomorrow.

That's our opportunity. And that is why AI governance will define the next generation of winners and losers.

About the Author

Eddie James Williams III is the Chief Security Officer of UNITI Cyber Solutions, a cybersecurity executive, AI governance strategist, speaker, and U.S. Army HUMINT veteran. With more than two decades of experience across military intelligence, healthcare security, consulting, government, and enterprise risk, Eddie helps leaders translate cyber and AI risk into clear business decisions. His work focuses on AI governance, GRC strategy, social engineering defense, cyber resilience, workforce development, and board-level communication. Eddie has been featured as a speaker with UMGC, BITCON, the Missouri Public Sector Cybersecurity Summit, and other cybersecurity forums, and has been recognized with an AI-focused Fluxx Award for innovation and thought leadership. He continues to write and speak on the human side of cybersecurity, cultural transformation, and responsible AI adoption.

Email: eddie.j.williams@uniticyber.com

LinkedIn: https://www.linkedin.com/in/cyberexecutive/

CHAPTER 31

THE NEW SCARCITY IS TASTE

By Pang Sern Yong
AI Strategist & Coach
Kuala Lumpur, Malaysia | Singapore

Most discussions around AI still focus on productivity and automation. How much faster can we write? How much cheaper can we design? How many workflows can we automate?

But I think a deeper shift is happening. AI does not just make work faster. It lowers the barrier to creation itself. When everyone can generate content, ideas, images, videos, designs, websites, strategies, and even music, output is no longer scarce. Creation becomes abundant. What becomes scarce is discernment.

The new divide is not simply between technical and non-technical people. It is between people who develop taste and people who are overwhelmed by infinite possibility. In this environment, human value shifts from execution to direction. The individual evolves from creator to curator, maker to editor, producer to signal finder.

Those who benefit most from AI may not be the fastest creators. They may be the ones who can identify what matters, filter noise from signal, and shape raw generation into something meaningful. The emerging superpower is not making more. It is choosing better.

The Moment We Are Living Through

The adult entertainment industry has always been one of the earliest adopters of technology — from DVDs, to 4K videos, to the internet. Generative AI is no different. Deepfake pornography has regulators scrambling, the public alarmed, and victims screaming for justice. This tells us something uncomfortable but important: when a technology lowers the cost of creation, it does not only create opportunity. It also creates chaos.

AI makes creation abundant. But the new advantage lies in knowing what to create, what to refine, and what to ignore. Right now, everyone can generate. But not everyone can create. With just a simple line of instruction — a prompt — things ranging from text, images, videos, posters, websites, music, and strategies can be created. The limit is no longer technical skill. The limit is imagination, judgement, and taste.

Before this, creation had friction. If you wanted a good video, you needed knowledge. You needed taste. You needed expertise. You might need to hire a VFX artist who could cost $10,000 and take a month to deliver. If you tried to DIY without the skills, things would often turn out badly. That barrier used to protect quality. It also excluded many people. AI has now broken that barrier. And because of that, we are seeing both the best and worst of human creativity at scale.

My Own Journey with Creation

I remember very clearly how difficult creation used to feel for me. I spent hours and hours writing, refining, rewriting, editing, asking people for feedback, and talking through my ideas. I had to be thick-skinned. I had to ask. I had to communicate. Sometimes it felt like a big ego trip, because every piece of feedback touched something personal. The same was true for images and videos.

I read countless books on writing. I attended workshops. I studied graphics, video, and film production. I even made my own film in 2018. It cost around $3,000, plenty of favours, and plenty of money. And honestly, one of the biggest barriers was not just money. It was favours.

I know what it feels like to need permission before I could create. I was not born into a rich family. I came from humble and modest beginnings. My parents were immigrants from Malaysia. I was not connected. I did not have networks handed to me. The only things I had were my hands, my mind, and my creativity. I was blessed to go through the world-class education system in Singapore, and I managed to build knowledge and expertise from the ground up. But none of it felt easy.

In my early years, I struggled a lot. I had to ask teachers for help. I had to consult peers. I watched with envy as other people seemed naturally talented — writing books, publishing academic papers, composing music, creating artworks, and winning awards. I remember one moment clearly.

In 2022, I was working with my writing coach, an English teacher from the US in her 50s. I remember waking up at 9am, walking to her door, gently knocking, worried that I might disturb her even though she had already said it was okay. From morning onward, we worked on just one paragraph of my bio. The next day, I still had to meet her again to continue writing. All I could offer in return was a meal. I remember asking another teacher who taught me screenwriting: "James, can you look at my writing?" Even after all that coaching, I never felt like a fully competent writer.

The same pattern happened in business. I grew my video marketing agency from the ground up in Singapore. Nobody taught me how to be an entrepreneur. I learned through self-study, through mistakes, and through operating in a competitive society where everyone was trying to get ahead. It was not always a sharing or supportive environment. It was often cut-throat. But that was also the kind of pressure that shaped Singapore and shaped me.

The Permanent Change

Then everything changed. The world permanently changed in November 2022 when ChatGPT came out. I was bewildered. I was overjoyed. I was impressed. I was amazed. Suddenly, everyone and anyone could become a creator. For someone who had spent years struggling to access knowledge, feedback, and creative support, it felt like many of my old barriers had collapsed overnight. Then came image generation. Then video. Then, shockingly, music.

What happened changed my life. I was a very early adopter. I discovered ChatGPT through TikTok doom-scrolling one day before it really went viral. That journey eventually led me to run a seven-figure agency focused on AI implementation and adoption for companies. I scaled from a solo business of more than a decade into a company with seven staff in Malaysia and regional operations. AI became a co-partner in everything. And it has been amazing.

But where I arrived was not accidental. It came from years of harsh life experience — travelling through Europe, Latin America, and Asia to 31 countries, visiting museums and galleries, watching films, learning culture, observing people, and trying to understand why some things feel good, true, beautiful, or meaningful while others feel empty. Over the years, I accumulated taste. And that became the game changer.

The Explosion of AI Slop

Within a few short years, AI had exploded everywhere. And with that explosion came something ugly: AI slop. Not bad because it was made by AI. Bad because nobody knew how to judge it. We became inundated with low-quality AI-generated content. Everywhere we turned, there were AI-generated posters, AI-generated books, AI-generated videos, AI-generated LinkedIn posts, AI-generated websites, AI-generated presentations, AI-generated "thought leadership." Sometimes it was not just bad. It was scary, because the content had proliferated so widely that people could not always tell whether something was AI-generated or human-made.

Companies jumped onto the AI bandwagon and implemented low-quality workflows. A 2025 MIT NANDA report found that just 5% of integrated AI pilots were extracting millions in value, while the vast majority remained stuck with no measurable P&L impact. This was a reminder that using AI is not the same as using AI well.

Professional platforms became full of generic AI content. Bookstores started seeing AI-generated books. Restaurants used AI-generated menus. People signed up for shady-looking services and gave money or personal data to vibe-coded websites built in minutes. This was disappointing because poor AI use damaged trust.

People became more sceptical of AI. Some swore they would not patronise businesses that used AI-generated images or menus. Others dismissed AI entirely because all they saw was low-quality output. But the problem was not AI itself. The technology is there to create good AI-assisted work. The missing ingredient is taste.

How I Learned to Use AI Differently

When AI first came out, I preached the democratisation of AI. At that time, many people were hesitant. Some were resistant. Some were scared. Some were sceptical. Within a few short years, AI had exploded everywhere. Some learned from me. Some did not.

But what I noticed was this: my way of using AI was drastically different from many others.

As a language graduate and someone deeply interested in language, culture, and norms, I could use AI at a higher level. I did not want to produce AI slop. I wanted to use AI to enhance my business, reduce labour, reduce cost, and help small businesses scale to a level that previously only big companies could reach. Within two years, my company won awards at both national and regional levels. That was a proud achievement for me. We built credibility with very few resources, an all-remote team, and a high level of AI adoption. But the key was this: we were not just making content.

We were making sure the content was good. It had to communicate clearly. It had to be aesthetically strong. It had to represent our brand. It had to be something we were not embarrassed to put into the world. We created email mailing lists, infographics, websites, and educational videos. We wanted to be the ones using AI well, not contributing to the flood of AI slop.

I encouraged my team to build taste. To observe more. To review more. To go out into the world and see things. To make judgements. To ask: is this good, or is this just easy? That question matters more than ever.

Taste Is the New Advantage

This brings us back to the main point. AI makes creation abundant. But abundance creates a new problem: how do we decide what is worth keeping? We are so obsessed with the magical quality of AI that we sometimes forget the real issue is how humans build a relationship with AI.

Many people misunderstand AI. Some are overly excited and accept everything AI produces. Others are overly sceptical and reject AI completely. Living in Asia, I can see this divide becoming very profound. Trust has fallen in many areas. Scepticism is rising. People do not know what to believe, what to use, what to reject, or what to ignore. So the real question now is: how do we cultivate taste? How do we form opinions and judgement? When should we accept what AI gives us, and when should we reject it? How do we distinguish the generic from the genuine? This feels like the photography-versus-painting debate all over again — not just high art versus low art, but the question of whether a machine-assisted medium can still carry human intention, authorship, and soul.

But taste is acquired. Taste is cultural. Taste comes from exposure, education, attention, and repeated judgement. Taste is not about being rich or elite. It is about trained attention — the habit of looking carefully, comparing deeply, and refusing the lazy option.

If you know what beauty looks like, you begin to develop taste. If you expose yourself to quality, you begin to notice quality. If you only consume cheap things, rushed things, generic things, and algorithmic things, your taste slowly gets trained downward. That is dangerous. Because in an AI-saturated world, even intelligent people can be fooled. People lose money to deepfake scams. People give their data to poorly built apps. People attend events that are basically ChatGPT-generated nonsense. People listen to talks that sound impressive but have no real point of view.

The signs are often there. But without judgement, people do not know how to reject.

What We Now Need to Judge

In this new environment, we need to ask sharper questions: What app is vibe-coded and unsafe? What person is real? What content is real? Which events are worth attending? Which talks are worth listening to? Which strategies are actually useful? Which designs are beautiful, and which only look impressive for three seconds? Which AI output should be accepted, refined, or thrown away completely? These questions are not technical questions only. They are human questions. They require consistency, alignment, quality, and taste.

How Individuals Can Survive AI Disruption

The most dangerous thing in the AI age is to accept without filtering. Accepting everything means giving away money, time, attention, trust, and sometimes even identity. At the same time, rejecting AI completely is also dangerous. It means refusing to learn a tool that is already changing how the world works. So individuals need a middle path. Do not compete with AI on volume. Compete on judgement. To survive AI disruption, individuals should focus on building the skills that AI does not automatically give them.

First, develop a point of view. Do not just ask AI to generate options. Ask yourself what you believe, what you stand for, what you want to say, and what should exist.

Second, refine your sense of taste. Read better books. Watch better films. Visit better spaces. Study good design. Listen to good conversations. Expose yourself to quality, even if some of it is not available in your own language or immediate environment.

Third, learn to direct tools instead of competing with them. AI is not just a machine that gives answers. It is a collaborator that needs direction. The better your direction, the better the output.

Fourth, practise rejection. Do not accept the first draft. Do not accept the generic answer. Do not accept something just because it looks polished. Ask whether it is true, useful, aligned, beautiful, and meaningful.

Fifth, build information literacy. Learn to verify. Learn to compare. Learn to sense when something feels off. In a world of deepfakes, scams, synthetic media, and infinite content, this is not optional.

These are very human skills. And we should not forget that human skills may become even more important than AI skills.

The Future Belongs to Better Choosers

Now everything has become abundant. The ability to make more is no longer enough.

The future may favour those who decide what should exist, not just those who can produce it.

The divide will get bigger. Low-quality AI slop will flood the world. But high-quality AI-assisted work — work guided by human taste, judgement, culture, and discernment — will stand out even more. Those who can leverage this will scale. They will do amazing things. They will become AI-enabled superhumans, ready for the next era.

But the superpower is not simply prompting faster. It is not generating more. It is knowing what is worth making. It is knowing what is worth keeping. It is knowing what deserves to exist.

In the age of infinite creation, the most valuable person is not the one who can produce endlessly, but the one who can recognise what is worth bringing into the world. In the age of abundance, taste becomes power. And discernment becomes the new human advantage.

About the Author

Pang Sern Yong is an AI transformation strategist, author, public speaker, and media voice helping organisations move beyond AI hype into practical, measurable adoption. With experience engaging over 9,000 professionals and advising C-suite, CEO, and Board-level audiences, he helps leaders and teams turn tools such as ChatGPT and Generative AI into real improvements in productivity, decision-making, communication, and workplace performance. Pang specialises in leadership alignment, workforce capability development, role-based AI workflows, and structured implementation support. He is also the founder of a local ChatGPT Community on Facebook, which has grown to more than 53,000 members. Known for his pragmatic, hands-on approach, Pang helps professionals adopt AI with clarity, discipline, and confidence.

Email: sernyong.pang@gmail.com

Website: http://theaihq.net

LinkedIn: https://www.linkedin.com/in/pangsernyong/

CHAPTER 32

BUILDING AI YOU CAN TRUST: LESSONS FROM A RISK-AVERSE INDUSTRIES

By Svetlana Zavelskaya, MS, MBA
Senior Technology Leader
New York, New York

No technology since nuclear fission will shape our collective future quite like artificial intelligence, so it's paramount AI systems are safe, secure, trustworthy, and socially responsible.

—Elham Tabassi

We're living through one of those rare moments when a technology arrives that everyone agrees will change everything. AI is the internet, the smartphone, the printing press, pick your historical parallel. The hype is deafening. The promises are intoxicating. And the pressure to adopt AI right now is immense.

But here's what very few want to talk about: Most organizations don't understand how to deploy AI responsibly. Many are terrified of being left behind, rushing to implement systems they don't fully understand, can't properly explain, and aren't sure they can control. The "fail fast" approach (IT slang for "Move fast and break things") gets applied to algorithms that make decisions about people's health, money, and lives.

This works great until something breaks. And in regulated, high-stakes industries, things breaking isn't just embarrassing. It's existential.

I've spent two decades in financial services and insurance. Both industries are highly regulated, both manage risk carefully, both are built on trust to the system and long-term customer relationships, both handle client assets, rely on data analytics and consumer confidence. Most people find this space boring until they desperately need it.

Insurance doesn't get to move fast and break things. We're a highly regulated industry where lawsuits are common and every decision affects real people's financial security. We're not known for innovation. We're known for being careful.

And that's exactly why insurance has something critical to tell the AI revolution: Here's how to build systems you can actually trust. Insurance is, at its core, the business of managing uncertainty. It's been doing it for centuries. When an insurance company deploys an AI system that incorrectly denies someone's claim or unfairly prices a policy, it faces lawsuits, regulatory investigations, and real harm to real people. The high stakes force discipline. This isn't theoretical for us. Every AI system we deploy has to survive this gauntlet.

That's our advantage: We've been forced to build AI that can actually be trusted, not because we're more ethical than other industries, but because we have no choice.

The Six Pillars of Trustworthy AI

Through years of working with AI/ML in the world's most regulated industries, I've stress-tested six pillars for building AI systems you can actually trust. These aren't philosophical ideals. They're practical necessities that have survived regulatory scrutiny, legal challenges, and real-world consequences.

Pillar 1: The Data Governance Foundation: You Can't Govern AI Without Governing Data

Here's an inconvenient truth that many organizations learn too late: You cannot build trustworthy AI on top of untrustworthy data. The most sophisticated model governance framework in the world cannot save you if your data quality is poor, your data lineage is opaque, or your data access controls are weak.

Financial services learned this lesson through painful incidents. AI models that were performing beautifully in testing would fail in production, not because the models were wrong, but because the training data contained systematic biases that testing didn't reveal. Credit models penalized zip codes that happened to correlate with protected characteristics. Fraud detectors learned patterns from biased historical enforcement. Risk models incorporated data that violated privacy policies.

The problem wasn't the AI. It was the data.

Why Data Governance Is AI Governance

The vast majority of AI governance failures have a data governance issue, and let me give you a few examples of how it happens:

- Bias in training data leads to the bias in models
- Uncontrolled data collection and processing results in privacy violations
- Undetected data distribution shifts cause model drifts

- Unknown data lineage brings explainability gaps
- Inadequate data access controls cause security breaches
- Uncatalogued sensitive data leads to regulatory violations

Poor data quality is a primary source of AI bias. Yet when organizations build AI governance programs, they often focus on models while leaving data governance as an afterthought. This is backwards.

The Data Governance Foundation

Data Quality Management: You cannot validate your AI if you don't know whether your data is accurate. Leading organizations assign data quality stewards and automate schema checks in their data pipelines. They set quality gates that prevent poor-quality data from reaching AI training pipelines.

Data Lineage and Provenance: Explainability requires knowing not just what your model does, but what data it learned from. When a model makes a decision that needs explanation, you need to trace it back through the model to the training data to the original source. Without lineage, you're operating blind.

Data Access Controls and Privacy: Different AI applications need different levels of data access. A customer service chatbot should not have access to the same data as a credit scoring model. Data access controls enable proportional governance through role-based access control, purpose limitation, data minimization, and privacy-enhancing technologies.

Data Cataloging and Discovery: You cannot inventory your AI if you cannot inventory your data. Modern data catalogs use AI itself to automate discovery, scanning databases, classifying sensitive information, suggesting business terms.

The Hard Truth

You can build the most sophisticated AI governance framework in the world, with clear roles, rigorous validation, comprehensive monitoring, and perfect documentation. But if your data governance is weak, your AI governance exists only on paper.

Organizations that get this right treat data governance not as a prerequisite to AI governance, but as an inseparable component of it. They build them together, resource them together, and measure them together.

Because at the end of the day, AI is mathematics applied to data. If you don't govern the data, you cannot govern the mathematics. And if you cannot govern mathematics, you cannot deploy AI you can trust.

Pillar 2: Explainability: If You Can't Explain It, You Can't Defend It

The first time you sit across from regulators demanding to understand an AI decision, you learn a hard lesson: "The algorithm said so" is not an acceptable answer.

In insurance, we must explain decisions to regulators, judges, and juries, especially when denying coverage or adjusting pricing. Traditional actuarial models are transparent; an actuary can walk you through variables and statistical reasoning. Modern AI systems, particularly deep learning, are often black boxes. They work brilliantly but can't always explain why. That's their power and their problem.

In regulated industries, unexplainable AI is unusable AI. We can't deploy systems we can't defend.

Our solution: Prioritize "glass box" AI, systems transparent by design. Sometimes this means decision trees over neural networks. Sometimes it means interpretability tools that decode black boxes. Sometimes it means hybrid approaches where AI generates insights that human experts validate and articulate.

We've learned to ask not just "Does this model work?" but "Can we explain exactly how it works and why?" Every industry deploying consequential AI needs the same discipline.

Pillar 3: Auditability: When the Paper Trail Matters

Explainability is about understanding a decision in the moment. Auditability is about reconstructing it months or years later.

Insurance claims can be disputed years after they're filed. Underwriting decisions can be challenged in court long after policies are issued. When that happens, we need to reconstruct exactly what our AI system knew, when it knew it, and why it made the decision it made.

This requires building audit trails into AI systems from day one, not as an afterthought. Every input that went into a decision. Every model version that was running. Every override or human intervention. Time-stamped, immutable, defensible.

We've seen companies implement impressive AI capabilities with no way to recreate what the system was "thinking" at any given moment. The model gets updated, training data gets refreshed, logs get overwritten. Good luck defending that decision in litigation.

In insurance, we treat AI decisions like financial transactions: logged, versioned, and auditable forever. We maintain model registries that track every version of every algorithm. We preserve training data sets. We document decision logic.

This isn't just defensive. It's how you learn. When an AI system makes a mistake, a robust audit trail lets you understand exactly what went wrong and fix it. Without that trail, you're just hoping it doesn't happen again.

Building auditability into AI systems adds complexity and cost. But it's not optional if you're serious about trust.

Pillar 4: Fairness: When Algorithms Inherit Human Bias

AI is very good at finding patterns in data, including patterns we wish weren't there. If your historical data shows that certain demographic groups filed more claims, your AI will learn that pattern. If your past underwriting decisions reflected human biases, conscious or unconscious, your AI will replicate them. Often with greater efficiency and scale than any human could.

The insurance industry has a complicated history with fairness. Redlining, discriminatory pricing, coverage denials based on protected characteristics—it happened. We're also one of the most heavily regulated industries specifically to prevent unfairness. We can't legally use race, religion, or certain other factors in our decisions. Full stop.

But AI doesn't understand "can't." It just finds correlations. And correlations can be proxies. AI isn't being intentionally discriminatory. It's being efficient. But the outcome is the same.

This is where insurance's regulatory framework becomes critical. We have to develop rigorous fairness testing for AI models and build fairness constraints directly into model training. This isn't about being politically correct. It's about avoiding massive legal liability and, frankly, doing the right thing. AI that perpetuates or amplifies bias isn't trustworthy.

Every organization deploying AI needs fairness testing as rigorous as their accuracy testing. In insurance, we've learned that "the model is just following the data" is not a defense. You're responsible for what your AI does, period.

Pillar 5: Accountability: Who's Responsible When AI Fails?

One of the most dangerous myths about AI is that it somehow dilutes human responsibility. "The algorithm decided" becomes a shield against accountability. In insurance, this doesn't fly. When an AI system makes a mistake—denies a legitimate claim or misprices a policy—the algorithm is not accountable. The company is.

This clarity is crucial for trust. Customers, regulators, and courts need to know who's responsible when things go wrong. And things will go wrong. All systems fail. AI systems fail. They make mistakes. They break in unexpected ways.

The question isn't whether your AI will fail. It's whether there's a clear accountability path when it does and a clear path to fix it.

Many companies establish explicit ownership for every AI system. There's a business owner who's accountable for outcomes. There's a technical owner responsible for model performance. There's a compliance officer who signs off on regulatory adherence. No ambiguity, no diffused responsibility.

There are also human-in-the-loop protocols for important decisions. AI can recommend, but humans approve. This isn't because we don't trust AI. It's because accountability requires a human in the chain. This slows things down compared to fully automated systems. That's the point. Speed without accountability is reckless. In regulated industries, we can't allow it.

Before you deploy an AI system, answer this question clearly: "When this fails, who's accountable?" If the answer is unclear, you're not ready to deploy.

Pillar 6: Resilience: Building Systems That Fail Safely

The final pillar is resilience, building AI systems that fail gracefully instead of catastrophically.

All systems fail. Software crashes. Models drift. Data quality degrades. The question is: What happens when your AI fails?

In insurance, we can't afford catastrophic AI failures. If a claims processing AI system goes down, we can't process claims until it's fixed. If a fraud detection system starts throwing false positives, we can't block all legitimate transactions while we retrain the model.

Resilience is built into AI systems through several principles:

Graceful Degradation: AI systems should fall back to simpler, more reliable methods when they encounter situations outside their training. Rather than failing completely or making confident mistakes, they should recognize their own limitations and escalate to human judgment.

Continuous Monitoring: Nobody can deploy AI and walk away. Teams monitor model performance in real-time, looking for drift, anomalies, and edge cases. When something looks wrong, they catch it quickly.

Kill Switches: Every AI system needs an emergency override. Not just the ability to turn it off, but the ability to instantly revert to a manual or simpler automated process.

Redundancy: Critical functions don't rely on a single AI implementation. Use ensemble approaches, human oversight, and backup systems so that no single point of failure brings everything down.

Building resilient AI costs more and takes longer than building fragile AI. But fragile AI isn't trustworthy, and in high-stakes environments, untrustworthy AI is worse than no AI.

The Blueprint for Other Industries

Insurance isn't unique in needing trustworthy AI. Healthcare makes diagnostic recommendations. Finance approves loans. Government determines benefits. Any domain where AI materially affects people's lives needs the same rigor.

The six pillars translate universally: Is your data correct? Are your AI decisions fair and accurate? Can you explain them? Can you reconstruct them months later? Who's accountable? Do they fail safely?

The insurance industry was forced to solve these problems first because of regulation and stakes. But every consequential AI deployment will face the same challenges. You don't have to figure it out from scratch. The blueprint exists. Build accuracy, explainability, auditability, fairness, accountability, and resilience into your AI systems from day one. Not as nice-to-haves, but as requirements.

Yes, this makes AI deployment slower and more expensive. But trustworthy AI isn't free. And untrustworthy AI is far more expensive in the long run: in legal liability, regulatory penalties, reputational damage, and human harm.

Conclusion

AI is transformative technology. But transformation without trust is just disruption, the dangerous kind.

Insurance's approach to AI won't win awards for speed or sexiness. We move deliberately, building things that last. Working for two decades in highly regulated industries, I'm convinced this is the right approach for any organization deploying AI where mistakes matter.

The future of AI isn't about what's technically possible. It's about what is responsibly deployable. It's about building systems people can trust, regulators can validate, and organizations can defend.

Insurance learned these lessons the hard way through regulatory scrutiny, legal challenges, and real consequences. We developed frameworks for trustworthy AI not because we're more ethical, but because we had no choice.

The rest of the world is about to face the same pressures. Regulation is coming. Lawsuits are coming. Public scrutiny is intensifying. Organizations that treat AI trust as an afterthought will pay the price.

You can learn from an industry that's managed risk for centuries: Build trust in from the beginning. Make data quality, explainability, auditability, fairness, accountability, and resilience non-negotiable requirements.

It's slower. It's harder. But it's how you build AI that works, not just technically, but in the messy, regulated, high-stakes real world where businesses actually operate.

That's the lesson from insurance: Trustworthy AI isn't about limiting what's possible. It's about ensuring what's possible is also ethical, explainable, and safe. In the age of AI, that might be the most innovative thing you can build.

About the Author

Svetlana Zavelskaya is a senior technology leader with extensive experience in the financial services and insurance sectors. She has spearheaded service delivery, enterprise architecture, engineering, cybersecurity, and operational management. A forward-thinking leader, she consistently delivers transformative outcomes by driving customer-centric innovation, operational rigor, and strategic technology adoption. She excels at building and scaling high-performing teams.

Svetlana's career is marked by a proven ability to orchestrate large-scale technology transformations, navigate organizational change, and implement cutting-edge solutions that align with business goals and generate measurable value.

Beyond her professional impact, Svetlana is a passionate mentor and active supporter of community service initiatives.

Email: svetlana.zavel@gmail.com

LinkedIn: https://www.linkedin.com/in/svetlana-zavelskaya/

CHAPTER 33

THE PERFECT EXCUSE: WHO IS ACCOUNTABLE FOR AI?

By Steven Yiliang Zhou
AI Transformation & Cybersecurity Expert
San Francisco, California

We shape our tools, and thereafter our tools shape us.
—Father John Culkin

In a slightly cramped glass-walled conference room, a small green light on a web conferencing microphone was flashing steadily in the center of a square table. Six senior executives from the client were seated around it, while I stood to the right of the projection screen, guiding everyone present to understand and accept the findings and conclusions being presented.

This was the final acceptance presentation I was delivering for this client. Over the past 10 minutes, I had used six pages to outline the management issues identified by my team using our custom-designed agentic AI, specifically, security operations cost issues that had been verified by the responsible parties. I felt it was time to move the presentation to the final decision-making stage, because I could read the room: They were leaning back in their chairs, nodding politely, arms crossed, all waiting for the meeting to end.

Then, the CFO at the corner of the table tapped the surface lightly with his right hand. After a few seconds of silence, he spoke, his tone half skeptical, half genuinely curious: "Why do you think your AI approach can identify issues so quickly? What if it's wrong?"

I looked around the table; everyone was staring at me. My voice remained steady as professional instinct took over: "The results have been validated against real data and confirmed by the responsible parties. The algorithms and models we used were fine-tuned based on your actual production environment. Furthermore, the data predicted by the AI is primarily intended for cross-checking and trend analysis. It is for reference only."

The CFO's attention drifted back to the budget breakdown on the screen. The meeting proceeded as usual, and no one pressed the issue further.

Hmm, for reference only.

Over the following months, those three words lingered in my mind. Not because they were wrong, but because they worked so well. Those three words made everyone in that room, myself included, nod in agreement and move forward, without having to worry about the real question that needed an answer: If the AI is wrong, who in this room takes responsibility?

AI doesn't just create new capabilities. It creates new vacuums, places where responsibility used to live and now doesn't.

I wasn't the only one who used this as an escape. In 2024, Air Canada's customer service chatbot comforted a passenger who had just suffered the loss of a loved one, promising a bereavement discount that didn't exist. The passenger sued the airline, which argued in court that its chatbot was an "independent legal entity" responsible for its own statements. Of course, the court rejected this defense. Yet the fact that such a major airline could so brazenly employ this line of reasoning speaks volumes about the critical juncture we've reached.

In this chapter, I aim to briefly outline the three AI fault lines I have observed: the fault line of verification, the fault line of accountability, and the fault line of purpose. These fault lines have existed since the moment organizations began deploying AI, yet most leaders have not been required to examine them.

After the briefing ended and everyone had left, the green light on the table continued to flash steadily. I didn't know if it was waiting for something or monitoring something.

The Verification Fault Line

When we adopt a technology, we assume we can test it and then trust the results it produces. For example, we stress-test bridges and audit spreadsheet formulas; we can trace the logic from input to output. If something goes wrong, you can pinpoint where and why.

AI doesn't work that way, at least not yet. And most leaders haven't truly grasped what that means.

The output of large language models is probabilistic. Their language is shaped by patterns derived from billions of data points. The results they produce may sound precise, authoritative, and comprehensive, but they could also be entirely fabricated. The output system lacks a truly robust mechanism for skepticism and verification.

In June 2023, New York attorneys Steven Schwartz and Peter LoDuca cited six fabricated cases generated by ChatGPT in a court filing. The judge ultimately imposed a $5,000 fine, stating that the lawyers' use of AI was not "inherently improper," but that ethical rules require lawyers to ensure the accuracy of submitted documents. This is merely a snapshot of how AI is currently applied in business, law, and other fields. AI tools present their outputs to professionals with the certainty typically reserved for true experts, yet there is often insufficient evidence to suggest that the output itself might be wrong.

Yes, every system makes mistakes, so AI does too. But that is not the crux of the matter. The crux lies in this: AI can make mistakes with confidence, fluency, and structural integrity. Its mode of error differs from that of other technologies, and those who rely on it may lack sufficient practical means to avoid the underlying harm. It is important to remember that traditional quality control asks, "Did it pass the test?" Whereas quality control for AI asks, "How do we know when not to trust it?"

Earlier, back in 2018, Amazon had already demonstrated another version of this flaw. It secretly used an AI recruitment tool that systematically downgraded resumes containing the words related to "woman." This happened because the AI had accurately learned that the company's historical hiring patterns favored men, and the AI simply optimized for that pattern. No one asked it to pursue fairness.

Validating AI is not a one-time checkpoint at deployment; it should be a relationship, one that is ongoing, skeptical, and resource-intensive. As long as the system remains in operation, this relationship must never be interrupted.

The Accountability Fault Line

AI differs from all previous flawed tools in one key way: The mistakes AI makes appear perfectly reasonable.

When a spreadsheet makes a mistake, it produces incorrect data. When machinery malfunctions, it leaves physical evidence. But AI systems generate outputs that carry the texture of human judgment, involving trade-offs, reasoning, and interpretability, and these outputs are wrapped in language that sounds like it has been carefully considered. This is why AI becomes the perfect excuse. Everyone in the chain of accountability can point to the AI system and say, "A rational person might have reached the same conclusion."

Or, "For reference only."

Yes. I said those three words to a CFO in a conference room who raised a valid objection. But since then, I've heard variations of those three words in every industry and every scenario. The phrasing differs, but the effect is similar. It stretches "accountability" so thin that no one actually bears it.

Analysts say the model made the decision; managers say the analyst ran the model; executives say it's permitted by policy; suppliers say the client configured it themselves. Every role has a plausible answer, and no one takes the initiative to accept responsibility.

Even before AI, automation was already creating accountability vacuums. The Boeing 737 MAX's MCAS system led to two crashes, claiming 346 lives. In the aftermath, Boeing blamed pilot training issues, regulators blamed Boeing's shortcuts in certification, and engineers blamed management for applying too much pressure. Yet the software system sat at the center of every explanation and in the vacuum of every role's responsibility.

Back in that conference room, the CFO's question was valid, and my answer was valid too. But between his question and my answer, something slipped through the fault lines, the very same thing that risks slipping through the fault lines in every organization running an AI system.

The Purpose Fault Line

Validation and accountability are oversight issues. But the third fault line is quieter and, in a sense, more dangerous. Because it doesn't look like a problem at all; it looks more like success.

When an organization deploys an AI system to optimize a process, the objective is clear: This new system must identify the most efficient path toward specific performance metrics. The company's dashboard turns green, quarterly financials improve significantly, and everyone moves forward with enthusiasm. Optimization has a direction. But that direction isn't necessarily the one the organization wants.

United Health Group was once exposed for having an AI system that was rejecting insurance claims at a rate exceeding the company's own policy standards. Internal documents revealed that the system had been set with explicit cost-reduction targets; the AI could accurately measure metrics and make autonomous decisions, bypassing coverage that should have provided reasonable protection. The AI served the targets, not the patients filing insurance claims.

YouTube's recommendation algorithm is nothing new. Its top-tier AI recommendation algorithm is designed to maximize individual viewing time. It has learned that emotionally charged content holds viewers' attention better; even if the system isn't intended to radicalize the audience, the AI has discovered that this approach yields better metrics.

When we discover that AI excels at something, we redesign processes around what it does best. The things the system cannot measure—judgment, context, ethical considerations—quietly lose their weight.

The rise of AI is turning us into tools, not in the dramatic sense of "replacement." For ordinary people, it's a quieter process. When an organization's goals begin to revolve around what AI can optimize, the stakeholders within it become mere inputs in someone else's equation.

Machines are no longer tools. People are becoming the tools.

This reversal does not manifest overtly, but unfolds bit by bit, step by step. Until one day, when the organization runs smoothly and all metrics are in the green, from a capital perspective, no one cares anymore what the original mission was.

This Is the Way

It's a pleasant evening in San Francisco. I spent the entire day chatting with investor friends about future opportunities for the AI startup I'm building. Now, as the sun dips below the horizon, I walk alone along the undulating road.

A Waymo drove by. No driver. No hesitation. No sound.

Then a second one.

"This is the way." The reflexive thought in my mind slipped quietly from my lips.

As the Waymo turned a corner and disappeared, I thought back to the empty driver's seat and the passenger in the backseat, their face illuminated by a phone. We all trust a system we can't fully explain.

I, who had spent the entire day discussing what AI can do, suddenly saw, amidst the city's twilight streetscape, what AI cannot do: answer for itself. I realized that at this moment, I no longer needed to ask what AI could do, but rather to ask myself: "When it cannot, who will answer?"

Bottom Line

These fault lines are what I see, and they are what you see as well. I do not yet have the capacity to provide a sufficiently robust framework for resolution, neither for you nor for myself. However, there are three questions I believe are worth asking honestly, and asking the right people, as they can distinguish between an "organization that governs AI" and an "organization governed by AI."

- *Verification Question:* If we need to confirm that this AI output is incorrect, what is the process and timeline for doing so? (If no one can provide a concrete answer, then the organization does not have a verification process; it is merely engaging in a "trust exercise.")
- *Accountability Question:* If this AI system were to cause serious problems or harm tomorrow, who in the organization would step in front of the cameras to explain what happened? (Not "the team," but a specific person, someone with a name. If such a person does not exist, the organization lacks accountability; it has a void of accountability.)
- *Purpose Question:* Does this AI deployment serve the organization's true mission, or has the mission quietly begun to serve the AI deployment? (If no one cares

about this question, then the organization lacks strategic alignment; it is merely "drifting.")

Before the rise of AI, the cybersecurity field learned this lesson the hard way. For years, cybersecurity was viewed as a technical function. It took a whole generation of data breaches for more organizations to realize that cybersecurity is a governance capability. From Equifax to SolarWinds, every major breach revealed the same pattern: The technical team had tools, but the organization lacked governance. Capability was present; accountability was not. AI accountability is following the same trajectory. The question is—will your organization learn this lesson before problems arise or after?

* * *

I still think back to that conference room: the sleek glass walls, the massive screen, and that green light that kept flickering.

It was there that I felt something about trust slip through the fault lines. Not because someone let go, but because no one had been assigned to catch it.

The next time your organization's AI system produces a confident, well-structured, and seemingly reasonable output, I suspect someone will echo the words I spoke in that room.

"For reference only."

It allows people to walk away gracefully, without having to answer for the unexplainable. But I no longer view those three words as a disclaimer. I'll treat them as a vacancy, a spot that should be filled with a name, a title, and a decision-maker.

The answer to that position is not AI. The answer is always a person. Make sure your organization knows who that person is.

About the Author

Steven Yiliang Zhou is a technology strategist and entrepreneur with over 15 years of experience at the intersection of enterprise security, AI, and organizational governance. He has held senior roles at a crypto company, Tencent, PwC, Deloitte, and Ernst & Young, advising C-suite executives across fintech, crypto, and Web3 sectors on security strategy, risk frameworks, and AI transformation.

A Gartner-honored speaker, published thought leader, and guest lecturer at Fudan University's MBA Masterclass, Steven has designed and deployed AI-driven systems for enterprise clients—and witnessed firsthand the governance gaps that emerge when organizations move faster than their accountability structures can follow. He is co-founder of the Hong Kong Web3 Cybersecurity Alliance and holds advanced certifications from ISC2, ISACA, and Forrester.

Today, Steven focuses on AI entrepreneurship, trend research, and enterprise AI transformation, helping organizations build the governance structures that make AI not just powerful, but accountable.

Email: zyiliang.us@outlook.com

Website: https://www.linkedin.com/in/yiliang-zhou-steven/

DID YOU ENJOY THIS BOOK?

If you enjoyed reading this book, you can help by suggesting it to someone else you think might like it, and **please leave a positive review** wherever you purchased it. This does a lot in helping others find the book. We thank you in advance for taking a few moments to do this.

THANK YOU

You might also like other Thin Leaf Press titles:

The AI Advantage: Thriving Within Civilization's Next Big Disruption

The AI Revolution: Thriving Within Civilization's Next Big Disruption

The AI Mindset: Thriving Within Civilization's Next Big Disruption

AI: Work Smarter and Live Better Within Civilization's Next Big Disruption

Peak Performance: Mindset Tools for Managers

Peak Performance: Mindset Tools for Sales

Peak Performance: Mindset Tools for Leaders

Peak Performance: Mindset Tools for Business

Peak Performance: Mindset Tools for Entrepreneurs

Peak Performance: Mindset Tools for Athletes

The Successful Mind: Tools to Living a Purposeful, Productive, and Happy Life

The Successful Body: Using Fitness, Nutrition, and Mindset to Live Better

The Successful Spirit: Top Performers Share Secrets to a Winning Mindset

Winning Mindset: Elite Strategies for Peak Performance

Winner's Mindset: Peak Performance Strategies for Success

The Life Coach's Tool Kit, Vol. 1

The Life Coach's Tool Kit, Vol. 2

The Life Coach's Tool Kit, Vol. 3

Ordinary to Extraordinary

The Magical Lightness of Being

Explore.

www.ingramcontent.com/pod-product-compliance
Lightning Source LLC
Chambersburg PA
CBHW071414230926
51846CB00001B/79
9781968318659